HYDRAULIC FLUIDS

**Peter Keith Brian Hodges, BSc.,
F.Inst.Pet.**
Petroleum Consultant
Norway

A
ARNOLD

A member of the Hodder Headline Group
LONDON • SYDNEY • AUCKLAND

Copublished in North, Central and South America by
John Wiley & Sons, Inc. New York–Toronto

First published in Great Britain in 1996 by
Arnold, a member of the Hodder Headline Group,
338 Euston Road, London NW1 3BH

Copublished in North, Central and South America by
John Wiley & Sons, Inc.,
605 Third Avenue,
New York, NY 10158–0012 USA.

British Library Cataloguing in Publication Data
A catalogue record for this book is available from the British Library

Library of Congress Cataloging-in-Publication Data
A catalog record for this book is available from the Library of Congress

ISBN 0 340 67652 3
ISBN 0 470 23617 5 (Wiley)

Typeset in 10/12pt New Century Schoolbook
by J&L Composition Ltd, Filey, North Yorkshire
Printed in Great Britain by J.W. Arrowsmith Ltd, Bristol
and bound by Hartnolls Ltd, Bodmin, Cornwall

CONTENTS

£43.85   £411
            9/9/
            1c 297

This book is due for return on or before the last date shown below.

0 9 FEB 1999

WITHDRAWN

621
.204
24

Hydraulic Fluids

PREFACE

This book is a revised edition of the original Norwegian language publication issued in 1994.

Today, hydraulics is an indispensable sector of modern engineering science. Enormous technological advances have been made since the initial use of water as a hydraulic medium in Joseph Bramah's press of 1795.

Despite the considerable number of publications dealing with hydraulics, the vast majority are principally concerned with the mechanical components and system design. Very few allot more than a chapter or so to the functional fluids which, after all, are the energy bearing media. In the following pages I therefore review the development of modern hydraulic fluids, discuss their physical/chemical properties in relation to operational requirements, and offer guidance concerning suitable maintenance routines.

It is my hope that this book may contribute to a wider understanding of the various fluid types and their discreet application.

I must admit to a sometimes overwhelming temptation to include additional data, documentation and discussion with respect to a number of my own particular fields of interest. Fortunately these urges were largely curbed by an exacting deadline, otherwise I would probably still be preparing a perhaps more lucid and comprehensive though unfinished text.

This foreword would not be complete without a sincere expression of appreciation to the many people who have assisted me during the preparation of the manuscript. Particular thanks to my previous employer, Shell Norway; also to publisher Birger Mølbach and Dag Viggo in Yrkesopplæring ans (Oslo) for invaluable assistance in printing the original illustrations. Last, but not least, I would express my gratitude for the encouragement and forebearance of my wife and friends during long periods dedicated to my PC alone.

Peter Hodges
Stabekk, Norway, 7th December 1995

GLOSSARY

Acid number
Total acid number (TAN) or neutralization value. Quantity of base, expressed in milligrams of potassium hydroxide, required to neutralize the acidic constituents in one gram of sample fluid.

Additive
Supplementary component modifying or improving fluid performance.

Adiabatic
Isentropic. Compression or expansion without heat being lost or taken up by the fluid.

Air entrainment
Dispersion of air bubbles into a circulating fluid, i.e. formation of an air-in-fluid emulsion.

Aromatic
Chemical compounds with a molecular structure incorporating the cyclic C_6 benzene molecule.

ASTM
American Society for Testing and Materials. A standardization association.

Base oil
A fluid, e.g. mineral oil or a synthetic fluid, without additives.

Bulk modulus
Reciprocal of compressibility, normally expressed in units of bar or megapascal.

Boundary lubrication
Lubrication of sliding contacts under conditions of high specific loading, resulting in the thickness of the lubricant film and surface roughness of the rubbing surfaces being approximately equal.

Cavitation
Hydrodynamic situation wherein vacuum cavities are formed momentarily and then collapse due to violent pressure changes. Usually accompanied

	by high noise level and frequently associated with erosive wear.
Centipoise	Unit of dynamic viscosity, 1 cP = 0.001 Pa s.
Centistoke	Unit of kinematic viscosity, 1 cSt = 1 mm²/s.
Coefficient of friction	Quotient of the normal load on a sliding surface and the force required to move the surface. May be determined as the static coefficient (μ_s) as movement just commences, or as the kinetic value (μ_k) under normal operating conditions.
Compressibility	Fractional volume reduction of a liquid when pressure is applied.
Density	Mass of unit volume of a substance, symbol ρ, expressed in units of kg/l or g/ml.
Demulsibility	Ability of a hydraulic fluid to separate from water.
Detergency	Ability to remove surface deposits displayed by certain polar fluids and additives. Detergent materials normally display a certain degree of dispersancy (refer below), and *vice versa*.
DIN	Deutsche Industrienorm – industrial testing and materials specifications issued by the German standardization association.
Dispersancy	The ability of certain fluids and additives to disperse other materials, contaminants, etc., in the form of minute particles throughout the base fluid.
Elastomer	A macromolecular material possessing elastic properties. It comprises of certain thermoplastic materials and vulcanized rubber, utilized for seals and flexible hoses. The commercial products are manufactured from various synthetic rubbers and polymers, modified by addition of fillers and other materials.
Emulsion	An intimate dispersion of one fluid within another.
EP-additives	Chemically active ('extreme-pressure') additives, generally based on sulphur

and phosphorous compounds, utilized to prevent catastrophic wear under heavily loaded boundary conditions. They function by reaction with the metal substrate, forming surface films which effectively prevent direct contact of the underlying asperities. Anti-wear additives of relatively moderate chemical activity are normally selected for use in hydraulic fluids.

Ferrography Laboratory technique for examining wear particles involving progressive separation of wear debris by passing the fluid through a magnetic field of varying density.

Flash point The lowest temperature at which the vapour above a fluid can be ignited under standardized test conditions.

Fluidity Inverse of viscosity, the flow properties of a fluid.

Friction Resistance to motion when attempting to slide one surface over another. Fluid friction is the internal friction of a liquid, i.e. the viscosity.

FZG Forschungsstelle für Zahnräder und Getriebebau, Munich. Gear test rig to evaluate anti-wear properties of lubricating fluids. Specification requirement in many hydraulic fluid specifications, e.g. DIN 51 524.

Helical molecules Molecules possessing a steric structure resembling a spiral spring, e.g. certain silicone fluids.

Homologues Chemical compounds possessing similar general structures, but different molecular weights. Typical examples are *propane, butane, pentane, hexane*, etc.

Hydraulic medium Hydraulic fluid, usually hydraulic oil. A liquid utilized to transmit hydraulic energy.

Hydrogen bond A strong secondary chemical bond (20–50 kJ/mol), and electrostatic intermolecular link between the hydrogen atoms in 'associated' liquids such as water and alcohols.

Hydrodynamics   Area of fluid mechanics pertaining to the behaviour of liquids in motion.

Hydrokinetics   Study of the energy of liquids in motion.

Hydrolytic stability   Ability to resist chemical reaction with water. Inferior hydrolytic stability can result in corrosion of susceptible metals and filter plugging.

Hydrostatics   Area of fluid mechanics pertaining to the energy of liquids under equilibrum conditions and under pressure.

Induction period   Initial period of time during oxidation of a fluid prior to an exponential increase in the oxidation rate.

Inhibitor   An additive preventing or retarding an undesirable effect, e.g. oxidation or corrosion.

Isentropic   See 'adiabatic'.

ISO   International Organization for Standardization

Isomers   Chemical substances of identical composition and molecular weight, but differing in molecular structure, e.g. *butane* and *isobutane*.

Isothermal   Compression or expansion at constant temperature, as opposed to the adiabatic process.

Laminar flow   Streamline flow conditions in a liquid, without turbulence.

Lubricity   Ability of a lubricant to reduce friction between mating surfaces under boundary conditions and moderate specific loads ('oiliness').

Neoprene   Polychloroprene (CR), synthetic rubber characterized by excellent ageing properties. Frequently applied as the external coating during the manufacture of hydraulic hoses.

Neutralization value   Quantity of base, expressed in milligrams of potassium hydroxide per gram sample, required to neutralize all acidic constituents in the fluid. Equivalent to alternative test methods reporting 'acid

	number', 'total acid number' or 'total acidity'.
Newtonian fluid	Liquids in which viscosity is independent of the shear rate.
Oxidation stability	Ability to withstand chemical reaction with oxygen/air and subsequent degradation. Of prime importance at elevated temperatures of operation.
PAG	Polyalkylene glycol (polyglycol), a class of synthetic fluids.
Particle analysis	Particle count, determination of the number and size distribution of solid contaminants in a fluid.
Pascal	SI unit for pressure, symbol Pa; 1 MPa≈10 bar.
Pascal's Law	Pressure applied to a confined liquid at rest is transmitted undiminished with equal intensity throughout the liquid.
Passivator	Type of additive preventing corrosion and the catalytic effect of metals on oxidation. Masks the normal electropotential of the metals by formation of surface films, e.g. sulphides and phosphates.
pH	Degree of acidity or alkalinity. The numerical value expresses the negative exponent of the hydrogen-ion concentration in an aqueous solution.
Polar substances	Molecules in which there exists a permanent separation of positive and negative charge, conferring a dipole moment to the molecule. Of significance for the adsorption of certain additives at metal surfaces, e.g. corrosion inhibitors and friction modifiers.
Polymer	Substance of high molecular weight formed by joining together ('polymerizing') a number of smaller units ('monomer') into large macromolecules. Typical polymers are the viscous polymethacrylate resins utilized as viscosity index improvers in hydraulic fluids.
Pour point	Lowest temperature at which a fluid will flow when tested under standardized test conditions.

ppm
Parts per million, e.g. mg/kg or ml/m^3.

Rate of shear
Velocity gradient within a fluid. In a fluid film between two sliding surfaces in relative motion, the rate of shear, expressed in reciprocal seconds, is equal to flow velocity divided by the thickness of the fluid film.

Reynolds number
A dimensionless value equivalent to the product of fluid velocity and pipe diameter divided by kinematic viscosity. The resulting value is used as a criterion to differentiate between laminar and turbulent flow conditions.

Scuffing
A serious wear mechanism involving microwelding of asperities on contacting surfaces under conditions of high pressure and high relative velocities. The microwelding is followed by rupture of the welds, roughening and increasing friction.

Seal compatibility
Ability of a hydraulic fluid and elastomer material to coexist in intimate contact without the elastomer displaying signs of undue swelling, hardening or deteriorating mechanical properties.

Specific heat capacity
Quantity of heat required to raise the temperature of unit mass of a substance by one degree. Usually expressed in kJ/kg per K or kcal/kg per °C.

Stick–slip
Jerky relative movement between sliding contacts under boundary conditions of contact. This phenomenon prevails when the static coefficient of friction is higher than the kinetic value. Addition of a friction modifier can alleviate the problem by ensuring $\mu_s/\mu_k < 1.0$.

TAN
Total acid number. The quantity of base, expressed in milligrams of potassium hydroxide per gram sample, required to neutralize all acidic constituents in the fluid. Equivalent test methods report the same property as 'neutralization value', 'total acidity' and 'acid number'.

Thermal conductivity	Ability to transmit heat, normally expressed in units of W/m per K.
Thermal stability	Measure of chemical stability when subjected to high temperatures, including resistance to molecular scission, i.e. 'cracking'. Regarding hydraulic fluids, this property is principally a criterion for the stability of additives.
Toxicity	Potential health hazards.
Vapour pressure	Measure of volatility, normally expressed in kPa, mm Hg or bar at a specified temperature.
Viscosity	Resistance of a liquid to flow when subject to a shear force; the internal friction of a liquid. See also 'centipoise' and 'centistoke'.
Volatility	Readiness to evaporate; the majority of non-aqueous hydraulic fluids have extremely low vapour pressures.
Viton	Elastomer based on fluorocarbon polymers (FPM). Compatible with most fluids up to ≈200°C and particularly well suited in connection with synthetic oils.
Wassergefährüngsklasse	WGK, the German classification system for assessing the potential toxicity of products in the event of pollution of waterways and lakes.
ZDTP or ZDDP	Abbreviation for the group of anti-wear additives based on various zinc dialkyl(aryl)dithiophosphate compounds. These additives also function, in varying degrees, as oxidation and corrosion inhibitors.

1

INTRODUCTION

1.1 Introduction

The word 'hydraulic' originates from the greek 'hydor' (water) and 'aulos' (pipe). The term 'hydraulics' is applied today to describe the transmission and control of forces and movement by means of a functional fluid. The relevant fluid mechanics theory concerns the study of liquids at rest (hydrostatics), or in motion in relation to confining surfaces or bodies (hydrodynamics). Hydraulic power transmission is the technique of transmitting energy by means of a liquid medium. Liquids utilized for this purpose are termed *hydraulic fluids*.

Use of hydraulics is expanding, and consumption of hydraulic fluids today constitutes a significant part of the world's total consumption of refined mineral oils, approximately 1 million tons per annum or around 10%. Mineral oil-based products represent over 90% of all hydraulic media; the remainder are various water-based fluids and synthetic oils. At present the bulk of these products are naturally utilized within the industrialized countries, but the demand for hydraulic fluids is now growing rapidly in the developing countries where vast future potential requirements exist.

Hydraulic fluids find innumerable applications in both static industry and mobile systems outdoors (transport equipment, excavators, bulldozers, etc.). Around 70–80% of the total volume of hydraulic fluids is utilized in static industrial installations. A certain amount of the remaining volume must meet the particularly critical quality requirements of specialized mobile systems in aerospace and military applications.

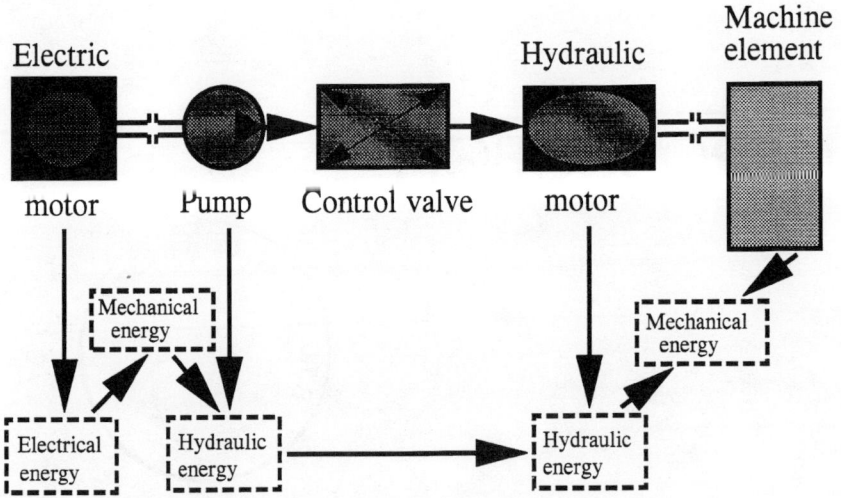

Fig. 1.1 Energy conversion in a hydrostatic system.

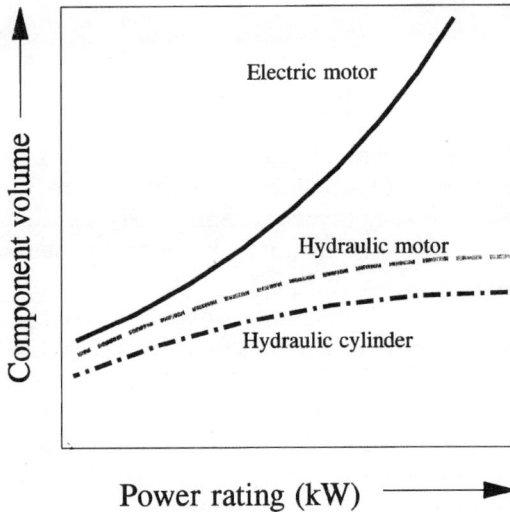

Fig. 1.2 Relative size of components.

The basic principle for hydraulic power transmission is illustrated in Fig. 1.1, where the input of electrical or thermal energy is converted to hydraulic energy, which is again transformed back to mechanical power for the output of the system.

Power transmission is effected by means of energy-converting units capable of transforming mechanical and hydraulic energy at the input and output of the installation. A functional fluid circulates in the hydraulic circuit, transporting energy between the input and output units.

One of the major advantages of hydraulic transmissions is the relatively moderate dimensions of the energy conversion units (hydraulic pumps and motors) compared to energy converters in other fields (Fig. 1.2).

The transmission of energy between fluid and conversion unit may be effected in accordance with *hydrostatic* or *hydrokinetic* principles (Figs 1.3 and 1.4).

1.1.1 Hydrostatic systems

Power transmission in a hydrostatic system is effected by means of the pressure of the hydraulic fluid, principal components being:

- the *hydraulic pump* to create the required working pressure;
- the *piping and flexible hoses* conveying the fluid flow between components;
- *valves* of various types controlling the direction of flow, pressure and volume;
- *cylinders* ('linear motors') converting fluid pressure to linear mechanical work, e.g. in a hydraulic press or to operate wing flaps on aircraft;
- *hydraulic motors* converting fluid pressure to rotary mechanical work, e.g. for the driving wheels of forestry machines or marine winches.

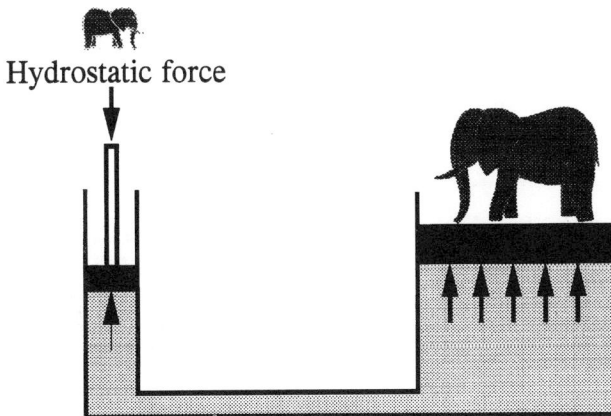

Fig. 1.3 A hydrostatic system.

1.1.2 Hydrodynamic systems

A hydrodynamic system consists in principle of a centrifugal pump ('impeller wheel') accelerating the transmission fluid against the inclined vanes of a turbine rotor (wheel). In the simple fluid coupling, kinetic energy is transferred from the circulating fluid and converted to rotary power as shown in Figs 1.4a.

Torque converters function in a similar manner. However, in this case a freely rotating stator is interposed between the impeller and turbine wheel – the stator is a vaned wheel. When the driven impeller rotates, the transmission fluid is circulated from its vanes to the turbine vanes, passing outside the stator. In returning from the turbine to the impeller, it passes through the stator. In doing so the stator's vanes accelerate the transmission fluid with increased energy via the impeller back to the turbine. The energy increase or output torque is greatest when the impeller is rotating at maximum speed and the turbine wheel is stationary. Figure 1.4b illustrates the principle of the torque converter..

Hydrostatic systems are the most widely applied form of hydraulics today, typical applications being hydraulic presses, machine tools, earth-moving machinery, hydraulic jacks and brake systems.

Hydrodynamic systems are used extensively for fluid couplings and automatic transmissions in commercial vehicles, agricultural machinery and passenger cars.

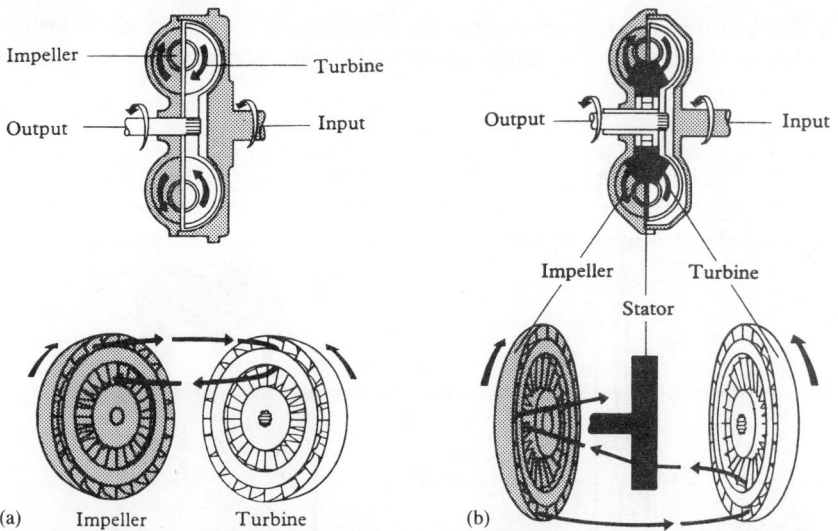

Fig. 1.4 (a) Fluid coupling (power transmission by kinetic energy). (b) Principle of torque converter.

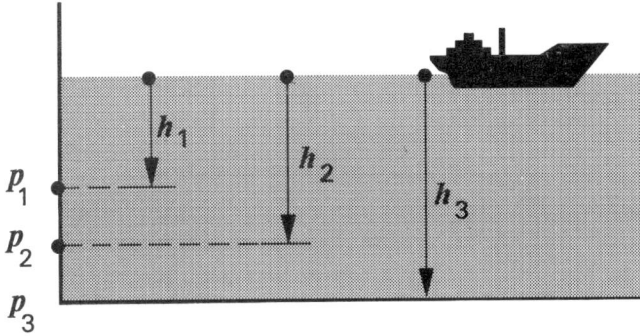

Fig. 1.5 Hydrostatic pressure.

1.2 Basic principles of hydraulics

In an open body of water, e.g. a lake, the static pressure at any point below the surface is proportional to the depth and density of the water: $p = h\rho g$ (refer Fig. 1.5). At 10 m depth in water, the resulting pressure is approximately 1 bar (100 kPa), whilst at the greatest depths of the west Pacific ocean, pressures of around 1140 bar (114 MPa) are recorded.

As long ago as ancient Egypt and Rome, a number of the fundamental laws concerning flow and equilibrum behaviour of liquids were utilized in the construction of simple water wheels, irrigation systems and water supplies. However, it was not before the 17th century that Blaise Pascal laid the foundation for the further development of hydraulic theory with his historic observation:

> The pressure at any point within a static liquid is identical in all directions, and pressure exerted on an enclosed liquid is transmitted undiminished in every direction, acting with equal force on equal areas. (Pascal's Law)

This condition is illustrated in Fig. 1.6, where a force F is impressed on the confined volume of fluid by means of a piston of cross-sectional area A. The resultant pressure is evenly distributed throughout the liquid and is equivalent to the unit load on the piston, i.e. F/A. Here we ignore the actual weight of the fluid which is normally of no significance on the pressure side of a hydrostatic system (a 10 m head of water is, for example, merely equivalent to 1 bar of pressure).

The fundamental principle for power transmission in a hydrostatic system is shown in Fig. 1.7.

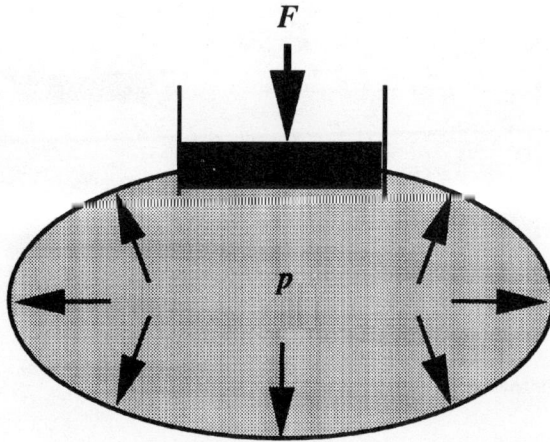

$$p \text{ (bar)} = \frac{F \text{ (newton)}}{A \text{ (cm}^2)} \text{ ; and 1 bar} = 9.81 \frac{N}{cm^2}$$

Fig. 1.6 Even distribution of pressure in an enclosed liquid.

The pressure developed in the fluid by force F_1 is F_1/A_1, and this pressure is transmitted unchanged (i.e. without pressure loss) to the piston of surface area A_2. The transmitted force F_2 is therefore identical to $F_1(A_2/A_1)$, and can lift a similar load or perform an equivalent quantity of other mechanical work. Assuming this force

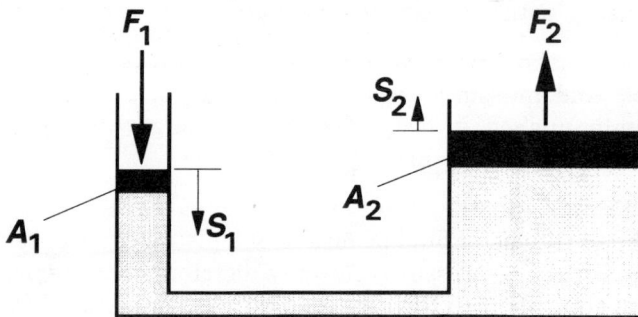

$$\frac{F_2}{F_1} = \frac{A_2}{A_1} = \frac{S_1}{S_2}$$

Fig. 1.7 Energy transmission in a hydrostatic system.

is sufficiently high to perform the task in question, then the linear movement of the two pistons is inversely proportional to their cross-sectional areas:

$$\frac{S_1}{S_2} = \frac{A_2}{A_1}$$

The volumetric displacement of the hydraulic medium is thus the decisive factor for the distance travelled by the individual pistons.

1.2.1 Pressure transmission

By rigidly joining together two pistons of different cross-sectional areas, the pressure developed in an adjacent hydraulic circuit may easily be increased or reduced (Fig. 1.8).

In the above figure, pressure p_1 acts upon area A_1 resulting in force F_1. On account of the pistons being joined rigidly together, the resultant force F_1 is transferred unchanged to the smaller piston of cross-sectional area A_2, i.e.

$$F_1 = F_2$$

and

$$p_1 A_1 = p_2 A_2$$

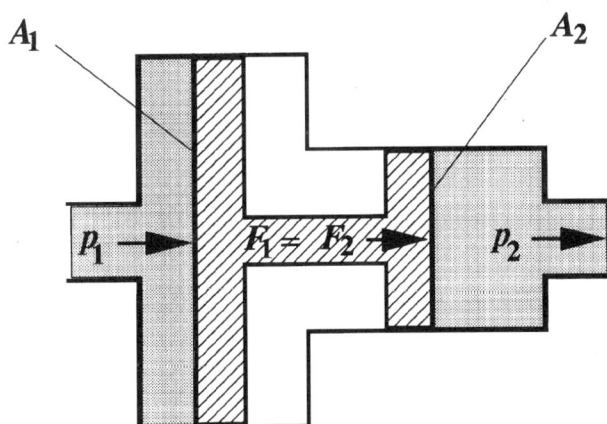

Fig. 1.8 Principle of pressure transmission.

or

$$p_1/p_2 = A_2/A_1.$$

By this simple arrangement, extremely large pressure changes may be achieved.

Example

	Piston #1	Piston #2
Diameter (mm)	60	15
Surface area (mm^2)	2828	177
Pressure (bar)	3	48

1.2.2 Liquids in motion

Hydrodynamics encompasses the mechanics of moving fluids. When a fluid flows through a pipe of varying diameter, the respective liquid volumes passing the different sections per unit time are identical (Fig. 1.9). Q = liquid volume flowing past cross-section A in time t. Hence the velocity of the fluid flow varies inversely proportionally to the cross-section of the pipe.

$$Q_1 = A_1v_1; \; Q_2 = A_2v_2; \; Q_3 = A_3v_3;$$

but $Q_1 = Q_2 = Q_3$ and therefore

$$A_1v_1 = A_2v_2 = A_3v_3,$$

which is known as the *continuity equation*.

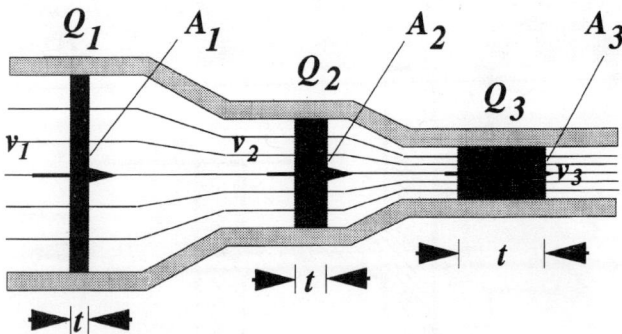

(Q = liquid volume flowing past cross-section A in time t)

Fig. 1.9 Variation of flow velocity.

1.3 Energy considerations

Bernoulli's law states that the total energy of a flowing, frictionless fluid remains unchanged provided no work is done by or on the fluid. In other words, the sum of pressure and kinetic energy remains constant, an increase in velocity being compensated for by a corresponding pressure reduction.

The total energy content (*W*) of a moving liquid is composed of:

(a) *potential energy*, related to the head of fluid,
(b) *pressure energy*, equivalent to the hydrostatic pressure,
(c) *kinetic energy*, related to fluid velocity.

Thus

$$W_{total} = p_{head} + p_{static} + E_{kinetic}$$

From the various relationships mentioned above, the following conclusions may be drawn:

- Increased flow velocity due to a smaller pipe diameter results in a higher kinetic energy.
- As the total energy content of the fluid remains unaltered, its potential energy, pressure energy or both must alter when the pipe diameter is reduced.
- In hydrostatic systems the pressure energy is the principal factor as the fluid head and velocity are relatively moderate.

Loss of energy by friction. When a liquid flows, a certain degree of friction occurs between the molecules of the fluid and all surfaces with which it is in contact. Consequently, part of the liquid's kinetic energy is converted to heat, resulting in a corresponding loss of pressure (Fig. 1.10).

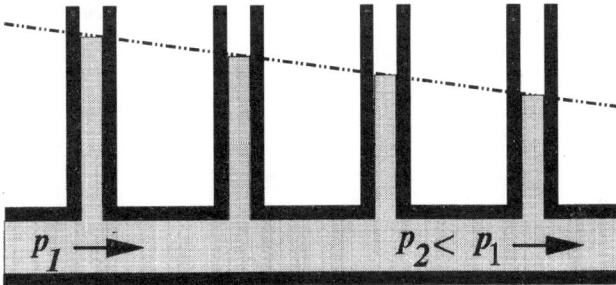

Fig. 1.10 Pressure drop due to friction losses.

Friction losses increase significantly when the flow pattern changes from laminar (streamline) to the turbulent form (Figs 1.11a and 1.11b).

The smooth, laminar flow pattern (Fig. 1.11a) is transformed to the unruly turbulent form when the combination of several parameters, e.g. viscosity, flow velocity, pipe diameter and surface roughness of pipe walls, exceeds a certain limiting value. For design purposes it is normal practice to utilize the dimensionless *Reynolds number* (Re) as a criterion to assess whether the flow is laminar or turbulent:

$$Re = v \cdot D \cdot 1000/v,$$

where v = flow velocity, D = pipe diameter (mm), v = kinematic viscosity (mm^2/s).

Laminar flow in normal, round, technically smooth hydraulic piping is usually stable up to a Reynolds value of Re $\leq$ 1200. Turbulent flow (Fig. 1.11b) with increasing friction losses will normally commence at Re $\approx$ 2300 (refer also to pressure-loss nomograms for hoses and pipes in Appendices 1 and 2).

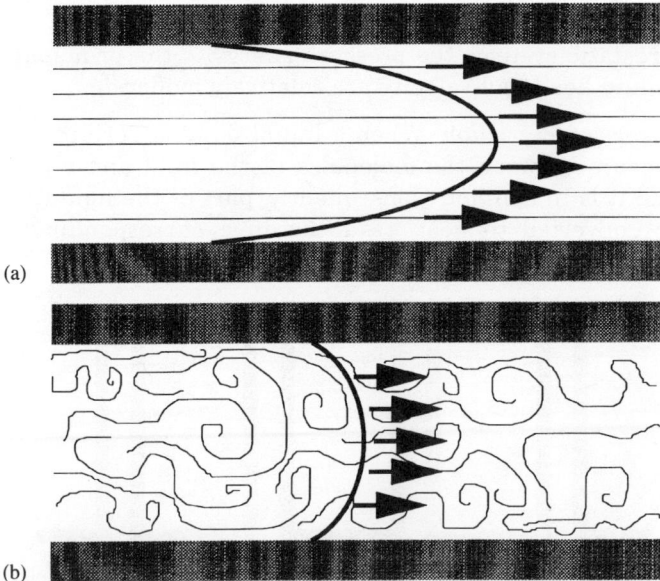

(a)

(b)

Fig. 1.11 (a) Laminar (streamline) flow. (b) Turbulent flow.

2

TYPES OF HYDRAULIC MEDIA

2.1 Historical

In the beginning there was water. . . . Thus it was also at the start of the hydraulic age (Fig. 2.1), and it seemed natural for Joseph Bramah to use water as the hydraulic medium for his historic press in 1795.

Water was also utilized as the hydraulic medium in the famous subterranean power transmission network which served a number of London's factories towards the end of the 19th century, and it was not until 1910 that the superior lubrication properties of mineral oils were required for the new pump designs of Armstrong, Williams & James. Nevertheless, it was the development of an oil-resistant, synthetic elastomer (nitrile rubber) during the 1930s that finally permitted construction of reliable hydraulic systems fully competitive with electrical and mechanical transmissions.

Continuing efforts to achieve improved efficiency resulted in designs incorporating higher operating pressures, but also higher system temperatures. Consequently, a requirement developed for fluids of higher quality, i.e. mineral oils displaying longer life in these new systems and providing better protection for hydraulic components under arduous operating conditions. Thus in 1940 the first inhibited oils were introduced, containing additives to counteract oxidative degradation and rusting.

The inherent advantages of hydraulic equipment were quickly appreciated by the mining and metallurgical industries, and introduction of hydraulic units proceeded with increasing momentum. These particular environments, however, posed extremely severe constraints

Water

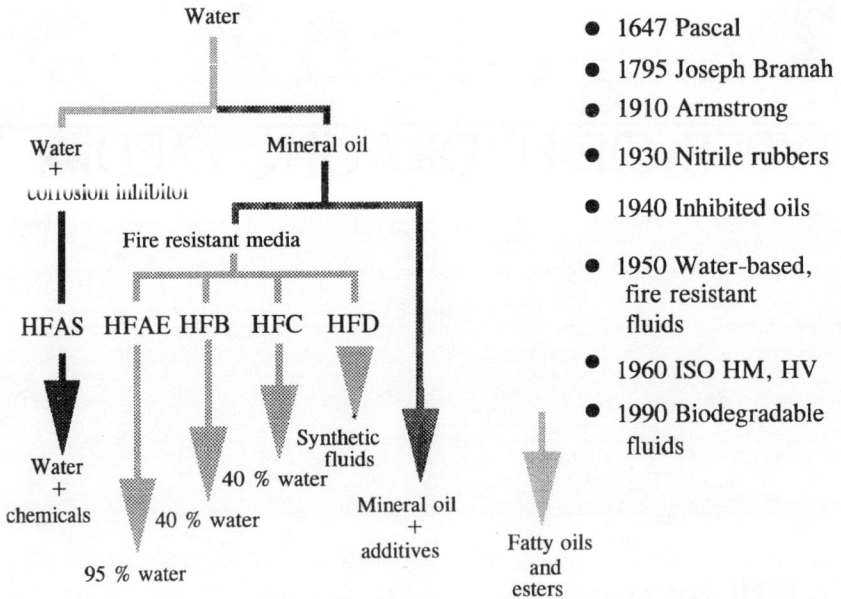

- 1647 Pascal
- 1795 Joseph Bramah
- 1910 Armstrong
- 1930 Nitrile rubbers
- 1940 Inhibited oils
- 1950 Water-based, fire resistant fluids
- 1960 ISO HM, HV
- 1990 Biodegradable fluids

Water + corrosion inhibitor

Mineral oil

Fire resistant media

HFAS HFAE HFB HFC HFD

Water + chemicals

95 % water

40 % water

40 % water

Synthetic fluids

Mineral oil + additives

Fatty oils and esters

Fig. 2.1 Development of hydraulic media.

with respect to fire resistance, and attention was immediately directed towards the obvious hazards connected with mineral oils at high operating pressures. This resulted in widespread research, culminating in the development of less flammable hydraulic media based on a variety of alternative materials.

New hydraulic components subsequently demanded fluids possessing considerably better anti-wear properties and the first products of this type (ISO type HM) emerged around 1960, followed by similar products with more favourable flow properties (ISO type HV).

During recent years increasingly greater attention has been directed towards health and safety considerations and the need to conserve our environment. Although *water* seems to represent the only truly environmentally friendly medium, the 1990s have seen an increasing availability of various biologically degradable fluids, white oils, etc., proposed as more environmentally acceptable alternatives to the conventional media.

Among the most used hydraulic media today are oil-in-water (o/w) emulsions, water–glycol solutions, mineral oils, esters and certain other synthetic fluids. Table 2.1 shows the classification of hydraulic media in accordance with ISO 6734 and DIN 51 502.

Table 2.1 Classification of hydraulic fluids in accordance with ISO 6734 and DIN 51 502

Description	ISO-L	DIN
Straight mineral oil (i.e. no additives)	HH	H
Type HH + oxidation/corrosion inhibitor	HL	H-L
Type HL + anti-wear	HM	H-LP
Type H-LP + detergent	—	HLP-D
Type HM + viscosity index improver	HV, HR	HVLP
Type HM + anti-stick/slip	HG	—
Synthetic fluids, non-fire resistant	HS	—
Oil-in-water emulsions (95% water)	HFAE	HS-A
Aqueous solutions of chemicals	HFAS	—
Water-in-oil emulsions (40% water)	HFB	HS-B
Aqueous polymer solutions (40% water)	HFC	HS-C
Synthetic fluids, phosphate esters	HFDR	HS-D
Synthetic fluids, chlorinated hydrocarbons	HFDS	HS-D
Synthetic fluids, blends of HFDR/HFDS	HFDT	HS-D
Other synthetic fluids, (non-aqueous)	HFDU	HS-D

A wide number of institutions issue recommendations, standards and specifications defining quality requirements for various types of hydraulic media and their applications:

- *component* manufacturers, e.g. Vickers, Rexroth, Denison,
- suppliers of *systems* for machine tools, off-highway vehicles etc. – *users* of hydraulic fluids,
- *government bodies*, where military specifications and environmental legislation is concerned,
- *national standardization committees*, e.g. the German DIN norms.

A number of specific quality requirements for the fluid in use are associated with the various types of hydraulic component:

- *Pumps/motors*
  Anti-wear
  Multi-metal compatibility
  Air release/anti-foam
  Demulsibility
  Low temperature flow properties
- *Valves*
  Corrosion inhibition
  Thermal and oxidation stability
- *Filters*
  Filterability
- *Seals/hoses*
  Elastomer compatibility

Table 2.2 Trends in the development of hydraulic media

Factor	Trend	Requirement
Increased efficiency	Higher pressures	Improved anti-wear properties
More compact systems	Higher temperatures	Better oxidation stability
Longer component life	Finer filtration	Good filterability
Reduced maintenance costs	Extended replacement intervals for components and hydraulic medium	Higher thermal and oxidation stability, improved anti-wear properties and finer filtration

The progressive development of new, more efficient systems creates requirements for hydraulic media of increasingly higher quality and technical performance (Table 2.2).

2.2 The ideal hydraulic medium

The ideal hydraulic medium is unfortunately non-existent, due to the conflicting nature of many otherwise desirable properties. Principal requirements are:

- satisfactory flow properties,
- a high viscosity index,
- low compressibility,
- good lubricating properties,
- low vapour pressure,
- compatibility with system materials,
- chemical stability,
- protection against corrosion,
- rapid air-release and demulsibility,
- good thermal conductivity,
- fire resistance,
- electrically insulating,
- environmentally acceptable.

Satisfactory flow properties are naturally of prime importance in a liquid transmitting forces from one location to another. This must apply throughout the entire range of temperatures under which the hydraulic fluid operates, not least during the initial start phase of the system under cold winter conditions.

A *high viscosity index* ensures comparatively moderate viscosity changes in relation to temperature fluctuations. By this means an appropriately wide temperature range for satisfactory operation of the system is achieved.

Low compressibility is advantageous as it ensures accurate transmission of pressure with minimum response time. Thus oscillatory motion and efficiency losses are minimized.

Good lubricating properties are a prerequisite for achieving acceptable service lives of components in modern high pressure hydraulic systems. There are also often special requirements with respect to specific frictional properties in order to ensure smooth, exact movements, e.g. in modern numerical control (NC) machine tools and robotics.

Low vapour pressure is desirable to obviate bubble formation or cavitation problems at the prevailing temperatures and low (possibly negative) pressures, at certain points in the system.

Compatibility with system materials is essential for a hydraulic medium. As operational parameters for hydraulic systems increase in severity, close cooperation between the manufacturers of new fluid types, seal materials, etc., is of vital importance.

Chemical stability is necessary to avoid disproportionately short replacement intervals for an expensive hydraulic fluid, and operational problems caused by degradation of products or deteriorating performance.

Corrosion protection is particularly important in hydraulic systems on account of the high pressures, fine tolerances and sensitive valves in modern systems. Contamination by condensation moisture is difficult to avoid in most systems. In general, hydraulic fluids should therefore contain effective corrosion inhibitors.

Rapid de-aeration and separation from water is necessary to maintain the specified performance level and counteract operational problems such as corrosion, cavitation, inaccurate pressure response, etc. These problems are reviewed in more detail in later chapters.

Good thermal conductivity is required to facilitate rapid dissipation of frictional heat generated in valves, pumps, motors and other components. Thus deterioration of the hydraulic medium or components is counteracted and a satisfactory efficiency rating maintained.

Fire resistance is an obvious advantage and is also subject to restrictive legislation for certain critical applications, e.g. in coal mines. The many types of fire-resistant medium are often inferior to conventional mineral oil-based hydraulic fluids in certain respects (see Chapter 20), and are usually more expensive.

Electrically insulating properties can be significant in a number of modern designs, e.g. oil well pumps, where electrical components are totally immersed in the hydraulic medium. Such instances demand close cooperation between component manufacturers and fluid suppliers to select suitable materials and the composition of the hydraulic medium.

Environmental acceptability covers many areas and often appears to be a misused characteristic. Concerning hydraulic media, the following factors are of principal interest:

- the working environment during handling and use,
- potentially injurious effects on surroundings should leakage occur,
- potential hazards during destruction or re-cycling.

On account of a superior combination of advantageous properties with respect to *compressibility, vapour pressure, lubricating properties, corrosion protection, chemical stability* and *price*, **mineral oil-based media** are predominantly the most common hydraulic fluids in use today.

3

MINERAL BASE OILS

The largest class of hydraulic fluids today is composed of refined hydrocarbon base oils, i.e. petroleum oils, containing suitable additives to improve and supplement the base oils' inherent properties.

Petroleum base oils are manufactured by a variety of refining processes from carefully selected crude oils, in many different viscosity grades. The purpose of these refining processes is to remove undesirable components from the original petroleum fractions, thereby optimalizing the chemical and physical properties of the raffinate. Crude oils from different geographic regions may vary substantially in chemical composition. The constitution of the refinery feedstock will strongly influence base oil performance, even though the effect can be moderated but not necessarily eliminated during processing.

3.1 Composition of mineral oils

Crude (petroleum) oil is an extremely complex mixture of numerous *hydrocarbon* compounds, i.e. substances composed of the chemical elements *hydrogen*, with chemical symbol H, and *carbon*, characterized by chemical symbol C. The crude oil also contains small quantities of sulphur, nitrogen, oxygen, vanadium, iron, nickel and other trace elements. The atoms are joined together in various sequences forming the individual molecular structures and classes of compound. On account of the many possible alternative structures as the number of atoms in a molecule increases, there are an enormous number of different hydrocarbon compounds present in a petroleum crude oil and most mineral base oils utilized for the production of hydraulic fluids. The general chemical and physical properties of the individual hydrocarbons are dependent upon the number of carbon and hydrogen atoms in the molecule and their geometric positioning, i.e. the steric structure.

3.2 Chemical nature

Due to their dissimilar atomic structure, carbon and hydrogen atoms display very different abilities to combine with other atoms. This property is termed *valency*.

Valency is a result of the electrostatic nature of the atom, principally derived from the electrons situated in the outer electron shell of the atom concerned. Carbon has a valency of 4 as opposed to 1 for

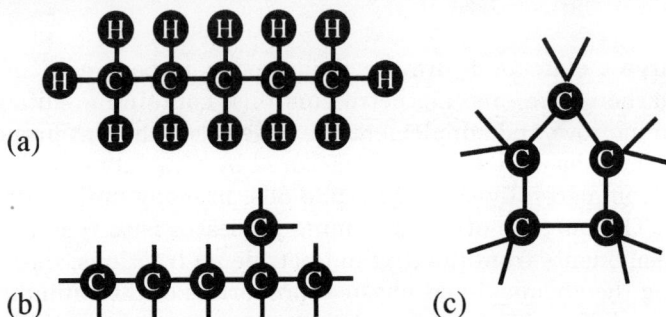

Fig. 3.1 (a) C_5 *normal*-paraffinic hydrocarbon. (b) C_6 *iso*-paraffinic hydrocarbon. (c) C_5 *cyclo*paraffinic ('naphthenic') hydrocarbon.

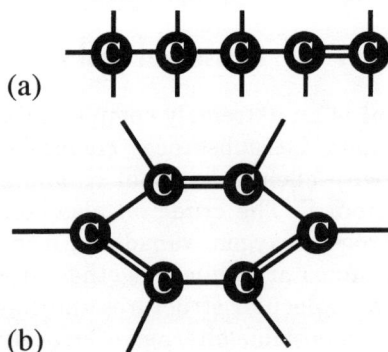

Fig 3.2 (a) C_5 olefinic hydrocarbon, (b) C_6 aromatic hydrocarbon.

hydrogen, and has therefore proportionately greater ability to combine with other atoms. On account of this multiple valency, the carbon atom is also able to link up with other carbon atoms, in the form of either *linear* or *cyclic* structures.

As shown in Fig. 3.1, the paraffinic hydrocarbons display a straight chain or branched configuration, unlike the cyclic form of the naphthenic compound, resulting in a considerable variation in properties. The *iso*-paraffins are characterized by side chains linked to the main linear hydrocarbon chain, and their properties also differ somewhat (considerably in certain respects) from the corresponding straight chain *normal*-homologues. (N.B. For simplification the constituent hydrogen atoms are only included in Fig. 3.1(a).)

When the carbon–carbon valency bonds are single bonds, the resulting molecule is termed *saturated*. Many hydrocarbons, however, include two or more carbon atoms joined by double or even triple carbon–carbon bonds. These *unsaturated* compounds display inferior chemical stabilty compared to the corresponding saturated compounds (Fig. 3.2).

Highly refined petroleum base oils are processed to remove all traces of *olefinic* compounds, these components being far too unstable for use in demanding applications such as hydraulic fluids due to their susceptibility to oxidation and subsequent deposition of lacquer-like deposits in the system. Olefins can nevertheless be formed during use in the event of the oil being subjected to temperatures above 320°C, for example in the near vicinity of over-dimensioned heating elements.

Although *aromatic* hydrocarbons appear to be highly unsaturated compounds, the alternating configuration of the double bonds (resonance) in the basic cyclic C_6 molecule of the simple mono- and di-aromatics results in surprisingly good chemical stability. During the refining of petroleum base oils by solvent extraction it is also apparent that a certain residual concentration of these aromatics is required to achieve optimal oxidation stability.

Mineral base oils contain molecules in many different sizes in each of the principal classes of hydrocarbon mentioned above, varying from simple, relatively small molecules to large, complex compounds. Mineral base oils for hydraulic fluids are normally composed of hydrocarbon molecules containing 20–50 carbon atoms, and have an average molecular weight (MMW) in the region of 350–550. All three of the principal hydrocarbon types are usually represented in the structure of the larger molecules and various methods are utilized to characterize the hydrocarbon type distribution of the complex base oil. Among the techniques employed are gas chromatography (GC), infrared absorption spectroscopy (IR), mass spectroscopy (MS), and in addition, various empirical methods based on physical test

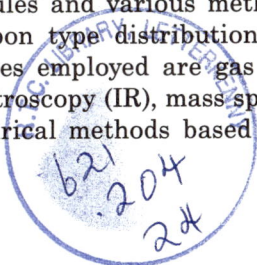

Table 3.1 Favourable properties of naphthenic and paraffinic base oils

Type of base oil	Advantageous properties
Paraffinic mineral oil (HVI)	Viscosity index Vapour pressure/volatility Viscosity: pressure coefficient Elastomer compatibility
Naphthenic mineral oil (MVIN)	Thermal stability Additive solubility Low temperature fluidity

Table 3.2 Typical test data for refined mineral base oils

Property	Test method	Naphthenic	Paraffinic
Viscosity (mm^2/s) at			
40°C (mm^2/s)	ASTM D445	8.3	27
100°C (mm^2/s)	ASTM D445	2.2	4.8
Kinematic viscosity index	ASTM D2270	49	97
Density at 15°C (kg/l)	ASTM D1298	0.87	0.87
Pour point (°C)	ASTM D97	<−45	−18
Flash point (°C)	ASTM D93	143	210
Initial boiling point (°C)	ASTM D447	278	370
Aniline point (°C)	ASTM D611	101	
Carbon distribution analysis (%m)	ASTM D3238		
C/A		2	6
C/N		48	26
C/P		50	68
Vapour pressure (bar at 100°C)	ASTM D2879	2×10^{-3}	3×10^{-4}

data. Despite the advanced physical–chemical analytical techniques available, supported by modern data processing, petroleum base oils nevertheless contain far too many isomeric compounds to be completely analysed down to the individual molecules.

Environmental considerations and technological advances have resulted in a trend away from the old, established refining processes, these being replaced or supplemented by catalytic hydrogen treatment under various conditions of time, temperature and pressure. The chemical stability of a base oil is just one of the important properties closely related to the refining processes and the essential requirement for a good hydraulic fluid is a base stock of high quality.

The major international oil companies possess considerable experience and expertise with respect to controlling the various operational parameters of these processes in order to achieve satisfactory yields and optimal quality from selected crude oil types. The companies are also engaged in extensive research programmes to develop new catalyst systems with increased efficiency for new processes.

Mineral hydraulic oils are usually based on highly refined *paraffinic* (HVI) oils, highly refined *naphthenic* (MVIN) oils or blends of both types. Each of these types of base oil possesses certain advantageous properties, mentioned in Table 1.3.

Provided feedstock and refining processes are carefully exploited, both of the above-mentioned base oil types will normally be characterized by good inherent oxidation stability, and their properties may be further enhanced by appopriate additive treatment.

A comparison of typical physical characteristics for highly refined naphthenic and paraffinic mineral base oils is illustrated in Table 3.2.

4
ADDITIVES

The final formulation of a hydraulic medium usually includes additives to enhance or supplement the base fluid's natural properties and achieve the predetermined performance characteristics. Prior to 1950, use of additives in hydraulic media (and lubricants in general) was relatively limited, interest being principally directed towards extending service life by means of oxidation inhibitors. More recently, the development of exacting high pressure hydraulic systems incorporating complex valve units at elevated operating temperatures has presented suppliers of hydraulic fluids with a vastly increased range of challenging quality requirements. Thus the development and test evaluation of a modern hydraulic fluid for the sophisticated equipment of today is often an intricate and expensive exercise, due to the numerous considerations in achieving a balanced, cost-effective formulation.

Many additives function better in certain base fluids than in others, and this additive response is of paramount importance in efforts to utilize additive components to their maximum advantage. Solubility characteristics of the base fluid are often decisive for the efficiency of many additives. Inadequate solubility may thus hinder an additive in fulfilling the envisaged function, whilst excessive solubility would be unfavourable with respect to, for example, the desired adsorption of surface-active agents on metal surfaces.

Hydraulic fluids intended for low or high temperature applications are often based on synthetic base oils with very different additive compatibility compared to mineral base stocks. Nevertheless, the manufacturing processes for synthetic oils usually permit the physical and performance properties of a specific fluid type to be adjusted to an optimum combination. Thus the solubility characteristics of, for example, polyalkylene glycols can be significantly modified by selecting other proportions or types of alkylene oxides during the manufacturing process. Synthetic hydrocarbon base fluids generally exhibit excellent additive response, but frequently require the addition of other components, e.g. esters, in order to obtain satisfactory solubility characteristics. Adoption of synthetic base fluids for high temperature applications gave increased impetus to research on additives suitable for use up to 200°C, particularly in various types of ester, and many effective additive combinations have been developed for this purpose.

Additives may also influence one another, either in a negative direction, or by displaying synergism and hence increased effectiveness. Selection of additive components and optimization of the chosen formulation may therefore be extremely time-consuming, but is nevertheless performed systematically by manufacturers dedicated to developing high-quality products. Along the way there is a progressive accumulation of know-how, an invaluable asset for later projects.

A simpler, though less reliable procedure, utilized by suppliers lacking extensive research and development facilities, is to purchase proprietary additive packages from independant suppliers and use these in whatever nominally suitable base oils are available.

The main types of additive used in hydraulic fluids are listed in Table 4.1.

Oxidation inhibitors are selected in accordance with the anticipated operating conditions, particularly the maximum temperature for which the product is designed. For high service temperatures, and in the presence of significant amounts of catalyst metal, additives of the metal passivator type are often required, e.g. metal dithiophosphates. An additional advantage when using many dithiophosphates is the improved anti-wear effect (see below).

Corrosion inhibitors include polar compounds, metal passivators, and substances capable of neutralizing acidic contaminants in the hydraulic fluid. Rust formation, especially, is a potential problem in hydraulic systems, on account of condensation moisture resulting from temperature fluctuations and moderate operating temperatures. Certain oil-soluble carboxylic acids possess the ability to form strongly adsorbed multi-molecular layers on ferrous surfaces, even at relatively low concentrations in the oil. These molecular layers are bound to the surface by powerful electrostatic forces and form an effective barrier against the penetration by oxygen and water that is necessary for rust-forming electrochemical reactions.

Defoamants utilized in hydraulic fluids today are normally silicone oils (high molecular weight polysiloxanes), present at a concentration

Table 4.1 Additives used in formulating hydraulic fluids

Type	Typical chemical nature	Typical %m
Oxidation inhibitor	Sterically hindered phenols, metal dithiophosphates, sulphurized olefins, aryl amines	0.2–1.5
Corrosion inhibitor	Carboxylic acids, benzotriazole, metal sulphonates, alkylated carboxylic acids	0.05–1.0
Defoamant	Polysiloxanes, organic esters	2-20 ppm
Anti-wear	Aryl phosphates, zinc dialkyldithiophosphates, organic sulphur/phosphorus compounds	0.5–2.0
Viscosity index improver	Polymethacrylate esters, styren/isopren copolymers, polyolefins	3–25
Pour point depressant	Polymethacrylate esters, naphthalene/wax condensation products	0.05–1.5
Friction modifier	Esters of fatty acids, fatty acids	0.1–0.75
Detergents	Metal salicylates, metal sulphonates	0.02–0.2
Seal swell	Organic esters, aromatics	1–5

of 2–20 ppm. Such silicones are practically insoluble in mineral oils, and are dispersed as minute spherical particles of diameter $\leq 1\ \mu m$ in the hydraulic fluid during the manufacturing process. Silicones are characterized by a relatively low surface tension, yet extremely high interfacial tension towards mineral oils. Due to this combination of physical properties, the surface active silicones can form small heterogenous areas within the bubble walls of surface foam. Due to their low surface tension, these small areas of silicone are weak spots and result in the bubble bursting.

It must be be remembered that anti-foamants of this nature can be deleterious under certain circumstances where significant amounts of false air gains access to the circulating fluid, e.g. at defective seals in a suction line. Under these conditions the surface-active additive will be adsorbed at the liquid:air interface whilst the bubble is still in the bulk fluid, and stabilize the resultant air-in-oil emulsion. This phenomenon is termed air-entrainment (see Chapter 11 describing aeration problems and Chapter 16 on test methods).

Anti-wear additives have been a subject of intense research during recent years. A principal target has been to develop universal additive systems of high thermal stability. Early formulations containing aryl phosphates yielded inadequate wear protection in more recent, demanding pump designs. More severe anti-wear requirements are currently specified by various national and component manufacturers against the general ISO HM classification. Good anti-wear performance is thereby ensured by critical combinations of realistic pump and rig tests, e.g. the Vickers 35VQ25 rotary pump and FZG gear test rig (see Chapter 16 reviewing test methods).

Zinc dialkyldithiophosphate compounds (ZDTP) containing secondary alcohol groups display excellent anti-wear performance for highly loaded steel–steel contacts. Unfortunately these chemically active *secondary* dithiophosphates also exhibit relatively poor thermal stability and were not particularly suitable for certain equipment incorporating components of bronze, e.g. some American Denison pumps. However, the thermal stability of the large dithiophosphate molecule can be considerably modified by the choice of appropriate combinations of the repective *alkyl or aryl* groups (Fig. 4.1). R, R', R'', and R''' represent optional alkyl or aryl groups derived from the respective *primary* or *secondary* alcohols utilized during synthesis of the additive. As is apparent from Table 4.2, thermal stability increases from additive A to additive C, whilst anti-wear performance decreases in the same direction.

Modern anti-wear hydraulic oils are therefore usually formulated with *primary* dithiophosphates, which permit formulation of oils possessing high thermal stability, yet at the same time achieving excellent results in the demanding European wear-test procedures.

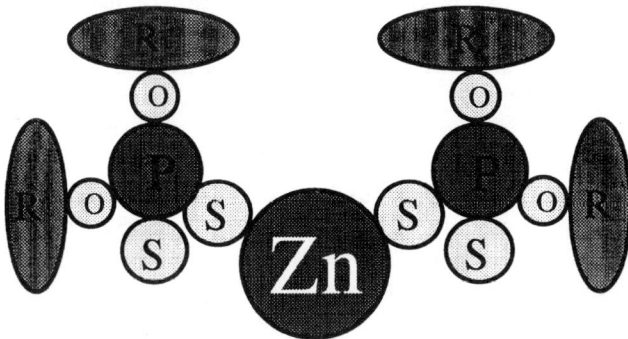

Fig. 4.1 Zinc dialkyl(aryl)dithiophosphate.

Table 4.2 Thermal stability and anti-wear properties of various zinc dithiophosphate additives

	A	B	C
Type of alcohol	Secondary alkyl	Primary alkyl	Aryl
Decomposition temperature (°C)	195	240	300
Shell 4-ball wear test, 1 h, 40 kg, 1500 rpm; wear scar diameter (mm)	0.41	0.48	0.99
Falex test, ASTM D 3233, failure load (kg)	431	409	340

An alternative means of achieving an acceptable combination of thermal stability and anti-wear behaviour was found in new, organic (ashless) sulphur/phosphorus compounds. Here again, there is a large range of additives to choose from, displaying varying degrees of anti-wear and thermal stability. The organic sulphur/phosphorus compounds are generally characterized by superior properties with respect to *hydrolytic stability* and *oxidation stability*, compared to the dithiophosphate additives.

Viscosity index improvers are normally high molecular linear polymers, often methacrylate resins, the manufacturing process being carefully controlled so as to yield products of relatively narrow molecular weight distribution. Good shear stability, i.e. resistance to mechanical degradation when subjected to high shear stresses (e.g. in a vane pump), is controlled by the chemical structure of the polymer and restricting the content of the highest molecular weight components.

The physical mechanism by which a viscosity index improver functions is essentially a solubility phenomenon. At low temperatures the large polymer molecules are tightly coiled together and may be likened to the individual peas in pea soup. As the temperature rises, the molecules gradually uncoil and assume a greater hydrodynamic volume, forming a network of polymer chains within the enveloping base oil (Fig. 4.2). By this means the polymer molecules exert a significantly greater restriction to viscous flow of the surrounding base oil at high temperatures (in other words, the additive displays a greater thickening effect on the base oil at high temperatures). As this is a solubility phenomenon, the efficiency of viscosity index improvers varies in different types of base oil. The various additives also influence the low temperature flow properties of a base oil differently, a factor of particular significance for outdoor use in cold climates.

When a polymer-thickened oil is subjected to high shear stresses, the polymer molecules tend to align themselves with the direction of

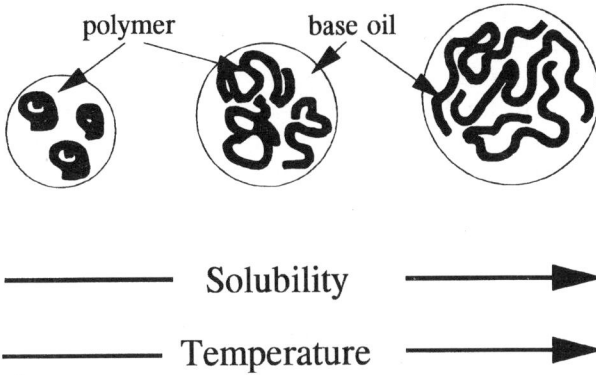

Fig. 4.2 Effect of temperature and solubility on polymeric viscosity index improver.

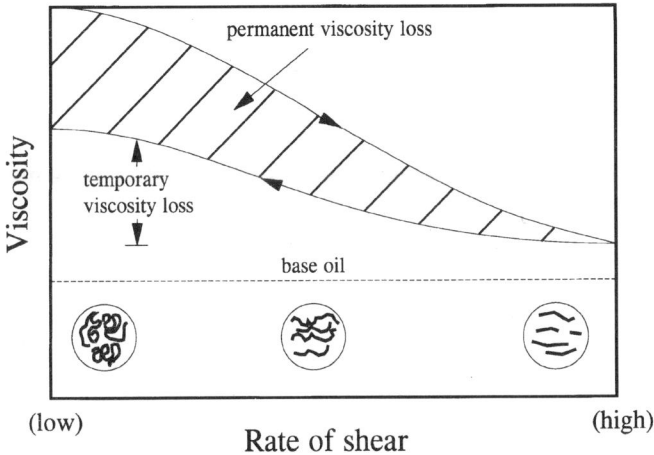

Fig. 4.3 Orientation and physical scission of polymeric viscosity index improver.

flow (Fig. 4.3). This will, for example, occur when the oil is forced through narrow clearances in hydraulic valves and pumps. The molecular orientation results in a temporary loss of viscosity, but the fluid reverts to its normal state as soon as the high shear stresses cease.

Exceptionally high shear stresses can also result in actual physical rupture of the polymer, the largest molecules being cut into smaller fragments. When this occurs, the fluid suffers a permanent reduction in viscosity, as the smaller polymer fragments are less effective thickeners. These temporary and permanent viscosity losses may be correlated with the type and concentration of the particular polymeric additive present in the base oil.

The *shear stability* of the polymer is consequently an important quality criterion for hydraulic oils subjected to high shear stresses during operation, due to the necessity of maintaining a certain safe minimum value of viscosity in order to ensure satisfactory lubrication.

Pour point depressants do not hinder the actual crystallization of dissolved wax, but function by modifying crystal growth and promote the formation of smaller crystals that do not interlock as easily. This is effected by the additive being adsorbed onto the surface of the developing crystals, thus forming an external barrier layer preventing mutual adhesion of the wax crystals and further growth into larger crystals. Hence a severely cooled, waxy oil containing pour point depressant additive resembles a homogenous slurry of minute wax particles dispersed in the oil phase.

Many low molecular weight methacrylate polymers are effective pour point depressant additives, and most methacrylate viscosity index improvers also display this useful secondary effect.

Friction modifiers are necessary under certain conditions to ensure smooth operation, free from juddering (stick/slip). These conditions include slow relative motion between heavily loaded mating surfaces, difficult combinations of materials, and exacting accuracy requirements, e.g. in numerically controlled machine tools. Selected friction modifiers have also successfully alleviated wear problems experienced in connection with polyurethane seals utilized in hydraulic cylinders.

As the relative speed between two lubricated surfaces in contact diminishes, the lubricant film thickness will also decrease, eventually resulting in increased physical contact and higher friction. Similarly when the same machine element is set in motion, the initial friction will also be high as surface asperities must be lifted over one another until sufficient speed is achieved to establish a continuous hydrodynamic lubricating film and separate the two surfaces. Without use of a friction modifier, there may be a serious risk of uneven operation under the conditions of speed, load and material combinations referred to above.

By addition of selected esters, fatty oils and/or fatty acids, it is possible to depress the static coefficient of friction (μ_s) below the kinetic value (μ_k) and satisfy the requirement

$$\mu_s/\mu_k < 1.0,$$

ensuring smooth, even motion between lubricated surfaces under boundary conditions.

Additives of this description are especially utilized in hydraulic media for numerically controlled machine tools, agricultural machinery incorporating wet brakes and automatic transmissions for vehicles (Fig. 4.4).

Detergents are polar substances providing a cleansing action with respect to surface deposits. These additives are used primarily in hydraulic media designed for high operating temperatures, or where significant contamination, for example by condensation moisture, is unavoidable. Detergent additives counteract the potentally injurious effect of lacquer deposits on valves, etc., in automatic transmissions, where high service temperatures are normal. A system subject to frequent contamination by water may also benefit from a balanced level of detergency, emulsified water being separated out of the oil phase during its comparatively long residence time in the fluid reservoir.

Elastomer swelling (seal swell) must be carefully considered when developing hydraulic media. Excessive swelling and softening cause wear or leakage due to displacement of the seal, while shrinkage results in leakage of the fluid. Most media are designed to permit a moderate degree of swelling without undue hardening of the elastomer material. Seal swell characteristics usually depend upon the base fluid and little can be done to remedy excessive swelling due to the inherent nature of a particular fluid. If, however, the base fluid shrinks the seal material, this may be corrected by addition of suitable seal swell additives such as esters, aromatics or ketones. These additives are particularly necessary for certain synthetic base stocks, e.g. polyalphaolefins, which exhibit a significant tendency to shrink the most common seal material of today, nitrile rubber (NBR).

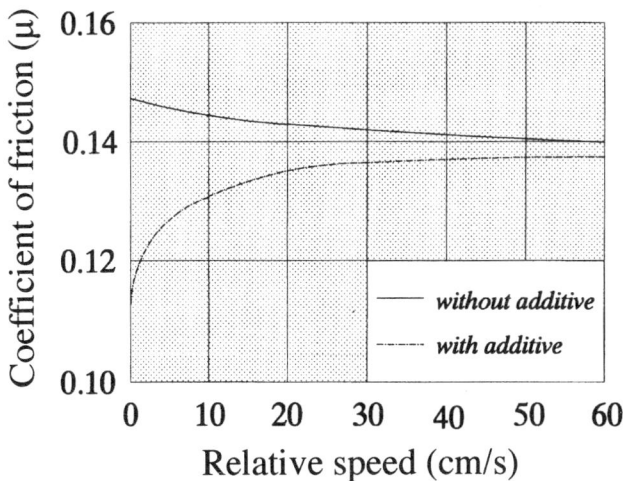

Fig. 4.4 Effect of friction modifier in automatic transmission fluid.

5

SYNTHETIC OILS

Synthetic oils are oligomers prepared by polymerization of selected monomers or similar oily fluids manufactured from other raw materials by chemical synthesis.

The wide variety of hydraulic applications pose a multitude of exacting physical and chemical requirements on the hydraulic fluid, in order to ensure efficient transmission of hydraulic energy under the prevailing conditions of operation. To what extent the individual fluids satisfy these demands is decisive for the normal area of application, temperature limitations, etc., of each individual fluid. Although mineral oil-based products still display the most satisfactory balance of properties for the vast majority of hydraulic applications, there are nevertheless many critical systems for which a specialized synthetic fluid may represent a preferable choice.

The principal reasons for selecting a synthetic fluid are:

(a) the best available mineral oil-based product lacks an essential property for the application in question;
(b) mineral oil-based products do not fully satisfy the specified operational requirements, e.g. adeqate oxidation life.

Certain fluid properties are inherently interrelated, and an excellent performance in one respect may frequently involve a corresponding limitation in another desirable property. A typical example of this is the somewhat mediocre frictional properties of some synthetic media otherwise possessing excellent oxidation stability. The explanation is that low friction under boundary conditions is strongly dependent upon adsorption and chemical reaction between the lubricant (hydraulic fluid) molecules and the surfaces in contact. Synthetic media of low polarity and high chemical stability are therefore relatively poor lubricants under boundary conditions. However, certain other synthetic fluids of high thermal and oxidation stability are normally the preferred choice for hydraulic systems operating at high temperatures (≥ 100 °C).

In general synthetic oils display superior performance, compared to mineral oil-based fluids, in one or more of the following respects:

- thermal stability,
- oxidation stability,
- viscosity:temperature properties (VI),
- low temperature fluidity,
- operational temperature limits,
- resistance to nuclear radiation,
- fire resistance.

According to their chemical composition, the individual synthetic media nevertheless possess greatly differing properties, and mineral oil-based fluids may be advantageous regarding:

- hydrolytic stability,
- corrosion protection,
- toxicity,
- compatibility with elastomers and constructional materials,
- solubility of additives,
- frictional characteristics,
- cost and availability.

5.1 Types of synthetic oil

Synthetic oils are normally classified in accordance with their chemical composition and Table 5.1 lists a number of the most important fluid types.

5.2 Synthetic hydrocarbons

Polyalphaolefins (PAO) are today one of the most well-known synthetic base oils, and are manufactured by polymerization of higher olefins (e.g. decene, C_{10}) followed by hydrogenation to remove the remaining double bonds and stabilize the polymer.

Commercial PAO is therefore characterized by an isoparaffinic structure, ensuring extremely favourable low temperature flow properties and a high viscosity index. Polyalphaolefins respond well to the

Table 5.1 Classification of synthetic oils by chemical composition

1. Synthetic hydrocarbons	(a) Polyolefins
	   polyisobutenes (PIB)
	   polyalphaolefins (PAO)
	(b) Hydrocracked mineral
	   oils (f.eks. Shell XHVI)

2. Polyethers	(a) Polyalkylene glycols (PAG) (b) Polyphenyl ethers (PPE)
3. Organic esters	(a) Diesters (b) Complex esters (c) Neopentyl polyolesters
4. Phosphate esters	(a) Aryl phosphates
5. Silicones	(a) Polysiloxanes (b) Silicate esters
6. Fluoroethers	(a) Fluoroalkylether (b) Perfluoropolyethers

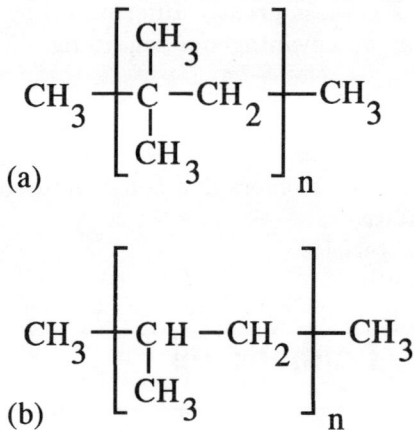

Fig. 5.1 (a) Poly-α-olefin, (b) polyisobutene.

addition of oxidation inhibitors and excellent oxidation stability may be achieved by careful selection of suitable additives. Consequently, on account of the elevated auto-ignition temperatures thereby attainable, polyalphaolefin base oils have proved suitable materials for the formulation of fire-resistant hydraulic fluids in certain aviation applications (e.g. military specification MIL-H-83282). Specialized polyalphaolefin-based hydraulic fluids have also found extensive use in critical systems on sub-sea oil installations.

Polyalphaolefins are commercially available in a wide range of viscosity grades, and even the lowest viscosity products display relatively low volatility up to 160°C in comparison with corresponding mineral oils. Polyalphaolefins also possess the advantage of unlimited miscibility with esters and mineral oils, thus simplifying conversion of a system from one product to the other. Although polyalphaolefins have much in common with highly refined mineral oils, they compare unfavourably in certain respects. The most pronounced weaknesses are their relatively poor solvent ability for many additive types and a tendency for low viscosity grades to cause shrinking of conventional nitrile rubber seals.

Polyisobutenes (PIB) show a similar isoparaffinic structure, but shorter side chains confer rather poorer low temperature properties compared to polyalphaolefins. Nor does polyisobutene display to the same degree the excellent oxidation and thermal stability of an inhibited polyalphaolefin. On the other hand, polyisobutene leaves minimal residual deposits following combustion or thermal degradation in the presence of oxygen. Consequently, polyisobutenes are extensively utilized in hydraulic systems within the aluminium industry, where contamination of the rolling oil by conventional mineral oils would result in unacceptable staining of the aluminium surface during the final annealing process.

Hydrocracked mineral oils are also composed essentially of isoparaffinic hydrocarbons, and resemble in many aspects the polyalphaolefins. However, with the exception of certain super-refined fractions, the molecular weight distribution of hydrocracked base oils normally spans a far wider range than corresponding viscosity grades of polyalphaolefins. Small amounts of heterogenous compounds are also present in the hydrocracked products. These differences result in generally inferior low temperature fluidity, but rather better lubricity and oxidation stability, compared to polyalphaolefins.

5.3 Polyethers

Polyalkylene glycols (PAG) or *polyglycols* are linear polymers prepared by a condensation reaction between selected alkylene oxides and glycols. By varying the proportions of, for example, ethylene and propylene oxide in the synthesis, it is possible to control the molecular weight of the product. It is thus possible to produce polyalkylene glycols in a wide range of viscosity grades and in water-soluble (hydrophile) or oil-soluble (oleophile) versions.

Many of the synthesized products possess a high viscosity index (KVI $\leq$ 220), good lubricating properties and favourable low and high

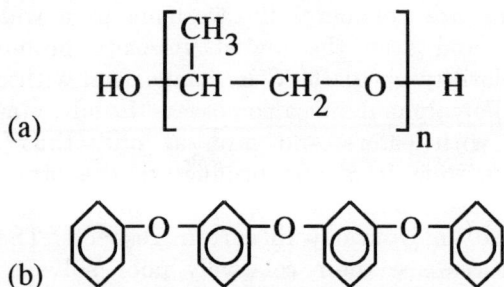

$$HO \left[\begin{matrix} CH_3 \\ | \\ CH \end{matrix} - CH_2 - O \right]_n H$$

(a)

(b)

Fig. 5.2 (a) Polypropylene glycol, (b) polyphenylether.

temperature performance. By use of suitable oxidation inhibitors, polyalkylene glycols achieve satisfactory oxidation lives at operational temperatures up to 150–200 °C, partly because their high solvent power enables significant amounts of oxidation products to be held in solution. Polyalkylene glycols are compatible with most constructional materials, but aluminium, magnesium, zinc and cadmium should preferably be avoided. The hygroscopic nature of polyalkylene glycols can in time result in the absorption of significant amounts of moisture, this aspect requiring particular attention to the corrosion-inhibiting properties of the hydraulic fluid formulation. Polyalkylene glycols are generally compatible with all usual seal materials, but attack many types of paint and only the solvent-resistant two-component epoxy products can be recommended for this application.

Water-soluble polyalkylene glycols find widespread use as base fluids for fire-resistant hydraulic media of ISO class HFC. Another important application is as a component for hydraulic brake fluid in the automotive sector.

Many polyalkylene glycols are biologically degradable, and this property has resulted in renewed interest in this type of hydraulic fluid as it is potentially less harmful towards the environment.

Polyphenylether (PPE) is regarded as one of the more exotic synthetic oils, not least on account of its high price (currently around 250 times the cost of a mineral oil). The cost factor, together with the extremely limited availability, has hitherto limited its use to specialized military and aerospace applications.

Polyphenylethers are, however, notable for their extremely high thermal, oxidation and chemical stability. Excellent lubricating properties, also under boundary conditions, low volatility, high stability against radiation and good hydrolytic stability amount to a unique combination of favourable properties for advanced technological applications. Apart from their high cost, the principal disadvantages of polyphenylethers are their disappointingly poor low temperature flow behaviour, negative viscosity indices and only moderate compatibility with elastomers.

5.4 Organic esters

Diesters are manufactured by chemical reaction between a dicarboxylic acid and a primary alcohol. Many esters are relatively easily biologically degradable, and this ability has resulted in a growing

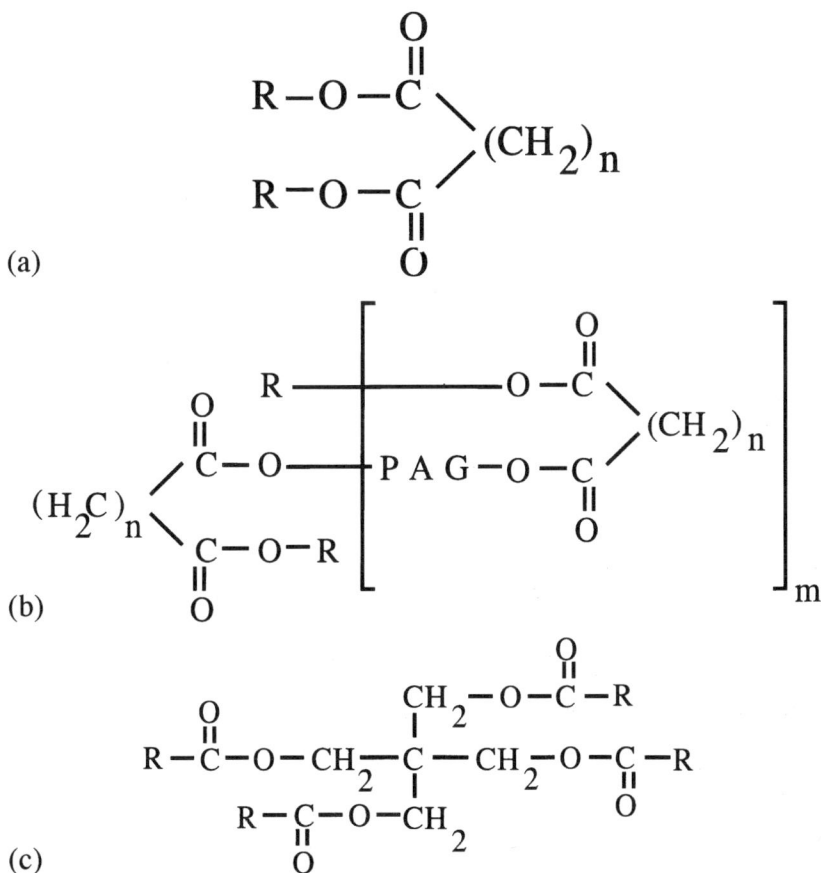

(a)

(b)

(c)

Fig. 5.3 (a) Diester (where R is an alkyl group, e.g. octyl C_8H_{17}, and n is typically 8–9). (b) Complex ester (where R is an alkyl group, e.g. octyl C_8H_{17}, and n is typically 8–9 and $m = 1.5$). (c) Neopentyl polyolester (where R is an alkyl group, e.g. octyl C_8H_{17}).

interest in the potential application of diesters as more environmentally acceptable hydraulic fluids.

Diesters display many favourable properties, e.g. excellent low temperature fluidity, high thermal and oxidation stability (after addition of suitable oxidation inhibitors), high viscosity index and good lubricating properties. Less favourable aspects are the poor hydrolytic stability and tendency to attack certain paints and elastomers associated with certain low molecular weight diesters.

Complex esters are also synthesized by reaction between a carboxylic acid and a primary alcohol, though in this instance a polyalkylene glycol is substituted for part of the alcohol component. This type of ester can be produced in far higher viscosity grades than the diesters, but does not possess the same excellent low temperature flow properties of the simpler diester, despite impressively low pour point test figures.

Complex esters exhibit better compatibility with elastomers than diesters and superior hydrolytic stability. Other properties are somewhat similar to the diesters.

Neopentyl polyolesters incorporate the unique symetrical 5-carbon neo-structure, conferring particularly good thermal stability. These esters also possess good oxidation stability, favourable low temperature flow characteristics and good lubricating properties. This combination of attractive properties has resulted in the selection of neopentyl polyolesters as the principal base stock for aviation gas turbine lubricants.

The price of these esters is normally higher than for diesters and complex esters, but the neopentyl polyolesters are potentially well suited for hydraulic applications at elevated temperatures.

5.5 Phosphate esters

Phosphate esters have been utilized for many years as fire-resistant hydraulic fluids in the industrial and aviation sectors.

The first commercial products were synthesized from coal-tar derivatives (cresols) and were consequently composed of a mixture of various isomeric aryl phosphates, including the neurotoxic ortho-tolyl phosphate. Today, other raw materials are used for the synthesis of phosphate esters; modern technology effects better control of the manufacturing process, and the toxicity of the final product is normally satisfactorily low.

The use of phosphate ester-based products in hydraulic applications is still principally dictated by fire-risk considerations. Although inhibited phosphate esters possess excellent oxidation stability and inher-

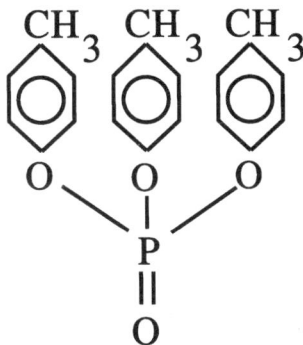

Fig. 5.4 Tritolyl phosphate.

ently good anti-wear properties under critical loading conditions, they nevertheless suffer from somewhat inferior hydrolytic stability, low viscosity index and extreme aggressiveness towards many conventional seal and coating materials. These weaknesses limit the use of phosphate ester-based fluids to specialized applications where a high degree of fire resistance is required.

Refer to Chapter 20 for a more detailed review of phosphate esters.

5.6 Silicones

The various types of siloxanes and silicate esters are frequently grouped together under the name **silicones**.

Polysiloxanes are semi-organic polymers and copolymers containing an inorganic backbone of repeating silicon–oxygen units and organic side chains substituted on the silicon atom along the polymer chain. The properties may be varied by selection of different substituents, though the most common are methyl and phenyl groups.

Dimethyl polysiloxanes have the highest viscosity index of all synthetic (and mineral) oils, also displaying extremely good thermal stability, oxidation stability, low vapour pressure and remarkably low pour points. They are, however, immiscible with most other fluids, including mineral oils, synthetic hydrocarbons, esters and polyphenyl ethers.

The ordinary unsubstituted polysiloxanes are notoriously poor lubricants for highly loaded steel–steel surfaces, but satisfactory with respect to steel–bronze contacts. The chlorinated polysiloxanes, however, generally yield satisfactory lubricating performance with heavily loaded steel:steel sliding contacts. Lubricity additives are also utilized, but choice is limited by solubility considerations.

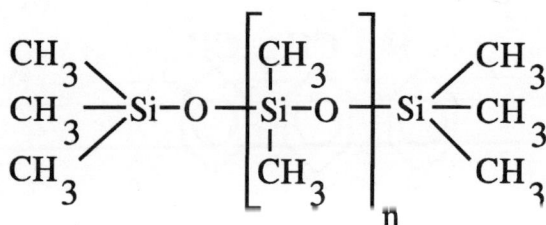

(a)

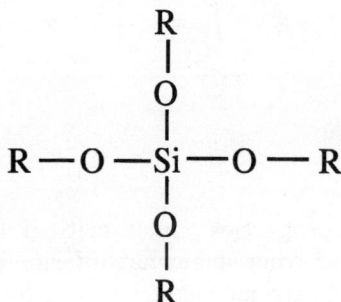

(b)

Fig. 5.5 (a) Dimethyl polysiloxane, (b) alkyl(aryl) ortho-silicate (where R is an alkyl or aryl group).

Despite a very low bulk modulus due to the helical molecular structure of polysiloxanes, these media have proved advantageous as functional fluids in certain servo and regulating systems, where their excellent viscosity–temperature properties are especially favourable. On account of their unusually high compressibility, the dimethyl siloxanes are frequently characterized as 'liquid springs' and utilized in various damping systems. Other hydraulic applications include certain shock absorbers and automotive brake fluids of type DOT 5 (see Chapter 21).

Silicate esters resemble polysiloxanes in many respects, and here again the physical and chemical properties of the fluid are determined primarily by the organic groups attached to the silicate entity. The properties of individual silicate esters may thus vary considerably. The base fluids are available in a range of viscosities, many displaying impressive viscosity–temperature characteristics, high thermal- and oxidation stability, and fair lubricating properties.

Although many silicate esters appear on first sight attractive base fluids for hydraulic applications, the relatively poor hydrolytic stability is often a serious constraint. Moisture in contact with the fluid represents a potential danger for hydrolysis and gel formation. Products based on a tetra-alkyl orthosilicate incorporating tertiary alkyl

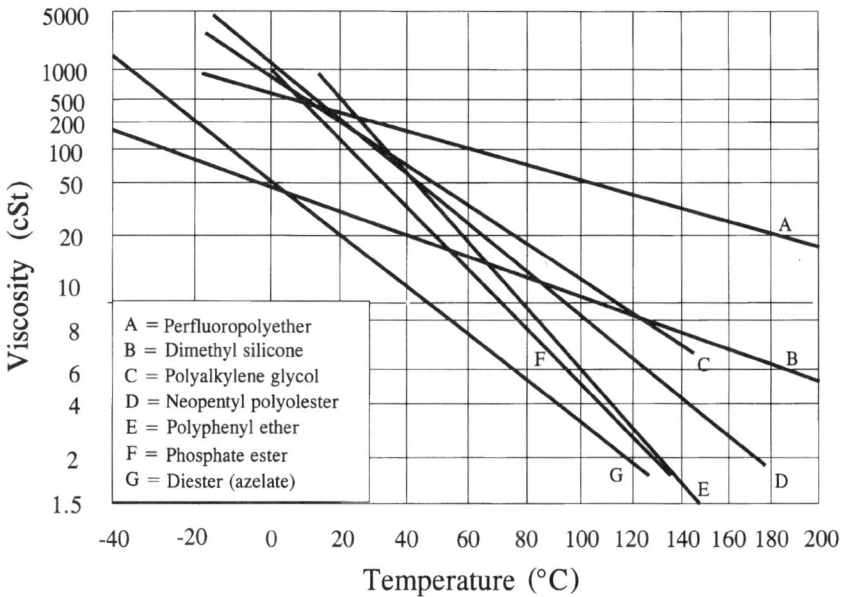

Fig. 5.6 Viscosity temperature relationship of various synthetic fluids.

Table 5.2 Typical physical and chemical properties of some synthetic oils[*]

	Poly-α-olefin	Polyglycol	Diester	Silicone	Silicate ester
Density at 15°C	0.80–0.85	0.95–1.0	0.90–1.0	0.90–1.1	0.85–0.95
Viscosity index	120–145	80–220	120–180	270–300	175–230
Pour point (°C)	−70/−35	−55/−20	−80/−30	−75/−40	−65/−50
Bulk modulus	High	Moderate	Moderate	Very low	Low
Oxidizes rapidly at, (°C)	170/230	150/200	150/200	200/260	180/250
Hydrolytic stability	+ +	− / +	− / +	− / +	−
Corrosion protection	+ +	− / +	−	− / +	− −
Lubricity/film strength	Good	Good	Good	Poor	Moderate
Volatility/vapour pressure	+	− / +	+ +	+ +	− / +
Solubility of additives	+	− / +	+	− −	−
Elastomer compatibility	+ +	+	− / +	+	+
Price (mineral oil = 1)	5	3–4	5	25–50	10

[*] Typical variation of properties within the same class is indicated where relevant.

groups (yielding a degree of steric hindrance) are the most stable materials and therefore generally preferable in this respect.

Silicate ester-based hydraulic fluids are used for aviation applications requiring temperature ranges of −65° to +204°C (military specification MIL-H-8446) and in the central hydraulic system of the supersonic Concorde aircraft.

5.7 Fluoroethers

Fluoroalkylethers and *perfluoropolyethers (PFPE)* are characterized by extreme chemical inertness and a high degree of oxidation stability, thermal stability and radiation resistance. In addition these fluids display good lubricating properties, low pour points, excellent fire resistance, compatibility with most plastics, elastomers and metals, and biological inertness. Negative aspects are high cost, relatively low bulk moduli, poor solvent ability for additives and immiscibility with other fluids.

The unique combination of favourable properties mentioned above has resulted in these fluorinated media finding application in a multitude of specialized applications within chemical and aerospace industries. Fluoroalkylether, perfluoropolyalkylether and perfluoroalkylether triazine are promising candidates for future development of fluids possessing an extreme degree of fire resistance.

The viscosity–temperature relationship of various synthetic fluids is illustrated in Fig. 5.6.

6

RHEOLOGY

6.1 Viscosity

The inner friction or resistance to flow of a fluid is of prime importance when selecting a hydraulic medium for a specific application.

The viscosity of a fluid, popularly designated 'thickness', expresses the inner friction of the liquid, in other words its resistance to flow. This property may be illustrated by imagining a number of golf balls being tipped ('poured') out of a bucket. The rubbing action of the balls as they roll over one another can be compared to the frictional forces between the molecules of a flowing liquid.

The degree of shear stress in a moving liquid is directly propor-

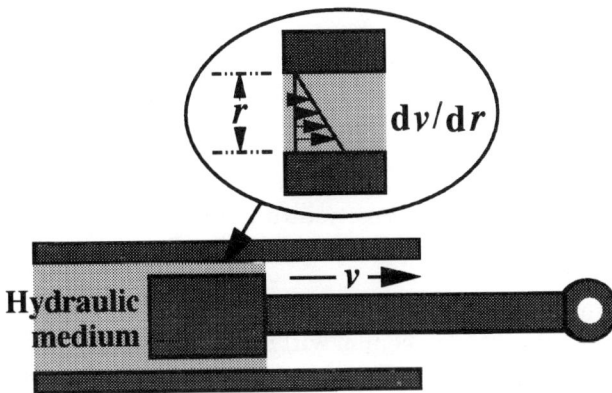

Fig. 6.1 Viscosity and rate of shear.

tional to the viscosity. A high viscosity gives rise to high shear stresses and consequently high resistance to relative motion of bodies separated by the fluid film. In 1687 Isaac Newton gave the first quantitative definition of viscosity, or 'lack of smoothness' as he called it, by means of the equation

$$\tau = \eta \cdot R_s,$$

where

τ = shear stress

η = dynamic viscosity

R_s = rate of shear.

Figure 6.1 depicts a cylindrical piston of peripheral surface area A and radial clearance r. The flow conditions within the film of liquid between the piston and cylinder wall may be illustrated by means of a pack of playing cards pressed down against a table top. If we move our hand forwards, the undermost card will remain stationary in its original position against the table, whilst the top card will follow the movement of our hand. All the cards in between will be displaced in relation to one another, as shown in the figure. The cards thus represent the numerous layers of liquid sliding upon one another, and it is this relative motion that constitutes liquid friction. This type of friction is considerably lower than ordinary dry friction between non-lubricated surfaces. Newton discovered that the force F required to move the piston, was proportional to the area A and piston velocity v, though inversely proportional to the radial clearance r. Hence:

$$F \propto A \cdot v/r$$

By inserting a proportionality constant (η), this relationship may be rewritten as

$$F = \eta \cdot A \cdot v/r,$$

where η is the *dynamic viscosity* in units of Ns/m^2 = Pa s. Dynamic viscosity is often expressed in *poise* (P) = 0.1 Pa·s, or *centipoise* (cP) = 0.001 Pa·s.

On account of the friction between the liquid and the cylinder wall, the outermost liquid surface film will remain stationary (as in the case of the pack of playing cards) whilst the liquid in contact with the piston will naturally move at the same speed as the piston (v). If we imagine the liquid in the radial clearance divided into a multitude of

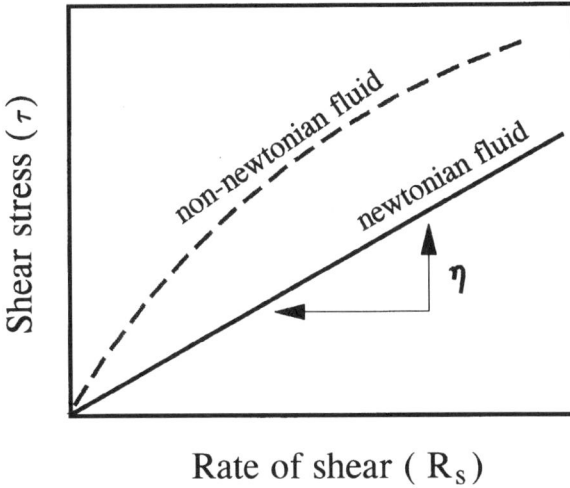

Fig. 6.2 Newtonian and non-Newtonian fluids.

thin, concentric layers, the velocity of the individual layers relative to the cylinder wall will vary from zero at the cylinder wall to v at the piston surface. The rate of shear (R_s) is dv/dr, the shear stress (τ) is F/A, and we can write

$$\text{Shear stress } \tau = \eta \, (dv/dr)$$

Liquids possessing a constant absolute viscosity at any given temperature, irrespective of the rate of shear, are termed *Newtonian* fluids (Fig. 6.2).

The majority of hydraulic fluids, with the exception of certain media containing high concentrations of polymeric additives and water-in-oil emulsions (HFB), can in practice be considered as Newtonian fluids.

The *kinematic viscosity* of a fluid in *Stokes* (St) or *centistokes* (cSt) is equivalent to the corresponding dynamic viscosity divided by density:

$$\text{poise (P)}/\rho = \text{stoke (St), in units of cm}^2/\text{s,}$$

$$\text{centistoke (cSt)} = 0.01 \text{ stoke, in units of mm}^2/\text{s.}$$

For hydraulic fluids (and industrial oils generally) it is usual to classify viscosity grades in accordance with the international (ISO) classification based on kinematic viscosity in centistokes (mm^2/s) at 40°C (see Table 6.1).

Table 6.1 ISO viscosity classification of industrial oils and fluids

Viscosity grade by ISO 3448	Kinematic viscosity cSt (mm²/s) at 40°C		
	Mean value	Minimum	Maximum
ISO VG 2	2.2	1.98	2.42
ISO VG 3	3.2	2.88	3.52
ISO VG 5	4.6	4.14	5.06
ISO VG 7	6.8	6.12	7.48
ISO VG 10	10	9.00	11.0
ISO VG 15	15	13.5	16.5
ISO VG 22	22	19.8	24.2
ISO VG 32	32	28.8	35.2
ISO VG 46	46	41.4	50.6
ISO VG 68	68	61.2	74.8
ISO VG 100	100	90	110
ISO VG 150	150	135	165
ISO VG 220	220	198	242
ISO VG 320	320	288	352
ISO VG 460	460	414	506
ISO VG 680	680	612	748
ISO VG 1000	1000	900	1100
ISO VG 1500	1500	1350	1650

This standard encompasses 18 main groups with the prefix ISO VG (viscosity grade), and is utilized by most suppliers of hydraulic fluids to classify their products. As an example Shell Tellus Oil 46 is an ISO VG46 fluid with a viscosity of 46 cSt±10% at 40°C. Viscosity classification of a hydraulic fluid in accordance with ISO 3468 does not, however, give any indication of the viscosity–temperature properties or of eventual non-Newtonian behaviour at high shear rates.

Energy losses in a hydraulic system must be kept as low as possible, and as long as the fluid flow is laminar, i.e. the fluid flows smoothly in parallel layers without turbulence, the thinnest fluids will achieve the highest hydraulic-mechanical efficiency ratings.

Use of thinner fluids will, however, incur increased leakage within the system past working clearances in pumps, valves, etc., reducing the volumetric efficiency of the system. In addition the lubricating properties of a fluid may become marginal as the viscosity is reduced to an excessively low value. Conversely, the application of a more viscous fluid will increase the volumetric efficiency of a system, but at the same time energy losses will increase due to higher fluid

Table 6.2 Typical viscosity limits for various hydraulic pump types

Type of pump	Viscosity (cSt)			Suction capacity mm Hg(bar)
	Min.	Recommended	Max.	
Gear				
plain bearings	25	25	1000	430 (0.57)
ball and roller bearings	16	20	1000	430 (0.57)
Piston				
seated valves	8	20	200	25 (0.03)
ports/sliding valves	16	25	500	125 (0.17)
Screw	25	75	500	480 (0.64)
Vane	13	25	850	250 (0.33)

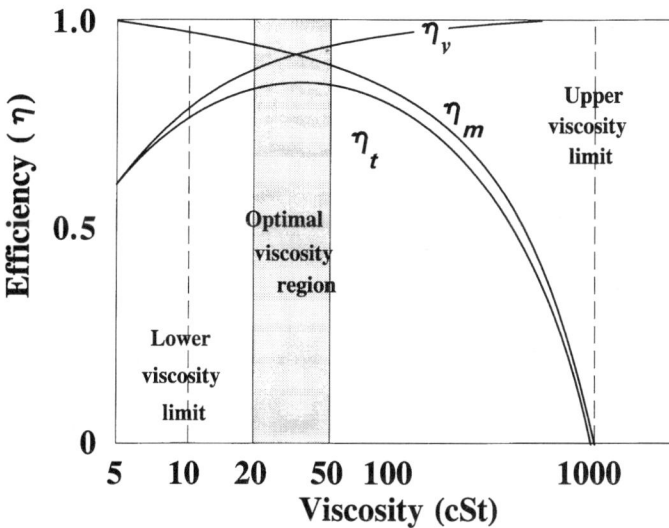

Fig. 6.3 Hydraulic efficiency as a function of viscosity; η_m, hydraulic-mechanical; η_v, volumetric; η_t, total.

friction in piping, filters, pumps and valves. The suction ability of pumps is normally the limiting factor for the maximum fluid viscosity tolerated by a hydraulic system. If friction losses within the system exceed the suction ability of the pump, the fluid will not completely fill the suction inlet and this will result in cavitation, i.e. formation of vapour bubbles and vacuum pockets. This phenomenon may often give rise to serious pump wear due to inadequate lubrication and erosive wear caused by the collapse of cavitation spaces.

(mm^2/s)

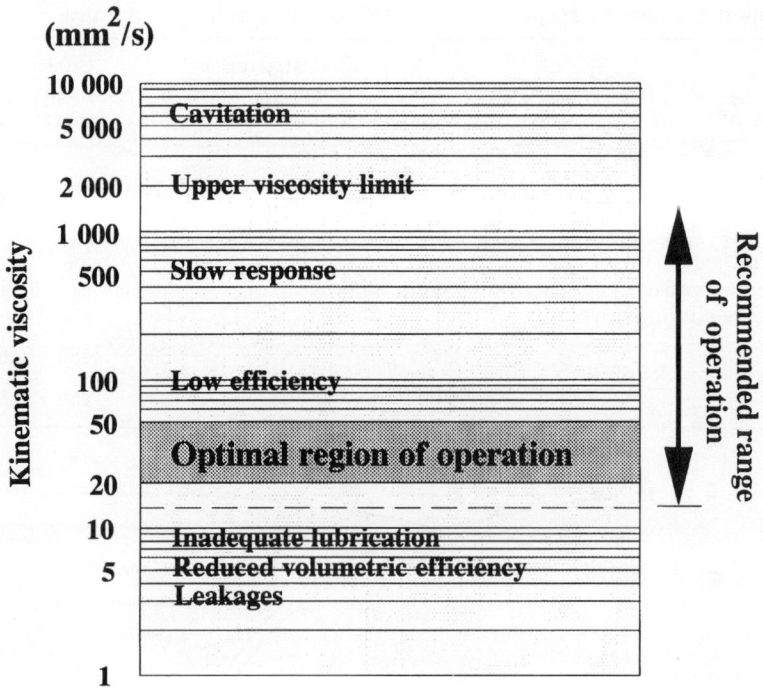

Fig. 6.4 Significance of viscosity for operation of a typical hydrostatic system.

The optimal viscosity for a hydraulic system is therefore a compromise between the lubrication requirements and the mechanical and volumetric efficiency (Fig 6.3). This balance of mechanical and volumetric considerations is best achieved by practical rig tests, and will often be found to lie nearest the limiting viscosity, ensuring adequate lubrication of the hydraulic pump. The various types of hydraulic pumps present dissimilar viscosity requirements, and winter conditions in, for example, Canada and the Nordic countries frequently represent a considerable challenge with respect to the selection of hydraulic fluids for satisfactory operation of outdoor equipment.

As a general rule, the viscosity of the selected fluid should be within the limiting values of 15–1500 mm^2/s (cSt) during starting and operation of the system, in order to avoid cavitation problems and undue wear of components (Fig. 6.4).

Although considered primarily to be a mechanical-related problem, cavitation effects on the suction side of the pump may also be promoted by restriction to flow caused by use of an excessively viscous fluid. The resistance or pressure loss due to the viscosity of the oil, coupled with the suction line pressure loss, should not create a

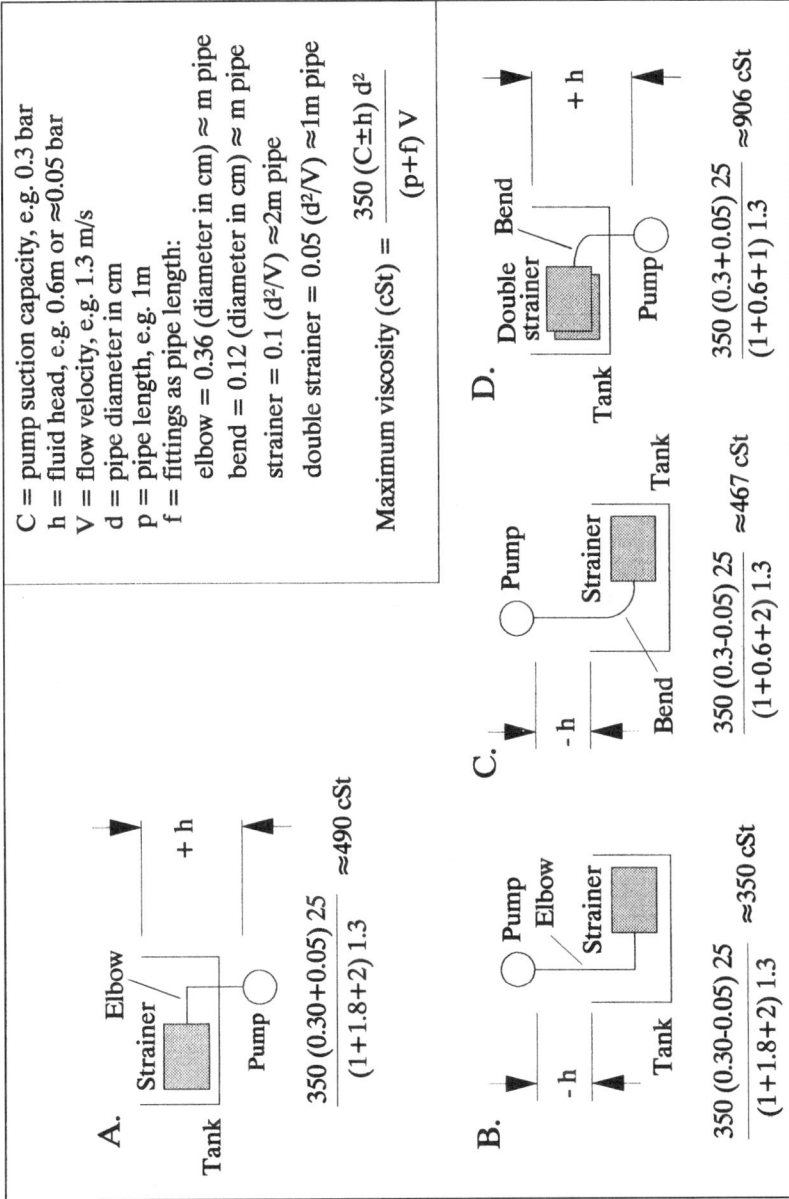

C = pump suction capacity, e.g. 0.3 bar
h = fluid head, e.g. 0.6m or ≈0.05 bar
V = flow velocity, e.g. 1.3 m/s
d = pipe diameter in cm
p = pipe length, e.g. 1m
f = fittings as pipe length:
    elbow = 0.36 (diameter in cm) ≈ m pipe
    bend = 0.12 (diameter in cm) ≈ m pipe
    strainer = 0.1 (d^2/V) ≈2m pipe
    double strainer = 0.05 (d^2/V) ≈1m pipe

$$\text{Maximum viscosity (cSt)} = \frac{350 \, (C \pm h) \, d^2}{(p+f) \, V}$$

A. Tank Strainer Elbow Pump + h

$$\frac{350 \, (0.30+0.05) \, 25}{(1+1.8+2) \, 1.3} \approx 490 \text{ cSt}$$

B. - h Pump Elbow Strainer Tank

$$\frac{350 \, (0.30-0.05) \, 25}{(1+1.8+2) \, 1.3} \approx 350 \text{ cSt}$$

C. - h Pump Strainer Bend Tank

$$\frac{350 \, (0.3-0.05) \, 25}{(1+0.6+2) \, 1.3} \approx 467 \text{ cSt}$$

D. Double Bend strainer Tank Pump + h

$$\frac{350 \, (0.3+0.05) \, 25}{(1+0.6+1) \, 1.3} \approx 906 \text{ cSt}$$

Fig. 6.5 Examples of maximum viscosity calculation.

suction depression beyond the capabilities of the particular type of pump concerned.

Cavitation in the suction system is a common cause of pump wear which can be very rapid. Since viscosity of the oil can be a contributory factor, a method of determining the viscosity at which the onset of this phenomenon can occur may be useful (Fig. 6.5). (N.B. The design parameters selected for these examples yield relatively restrictive values for the maximum allowable viscosity, compared to the general recommendations in Fig. 6.4.)

6.2 Low temperature flow properties

It is essential for hydraulic fluids intended for outdoor use in mobile equipment to possess satisfactory pumpability at the prevailing ambient temperature during the initial period of operation with cold fluid. Unfortunate starting procedures using too viscous a hydraulic fluid may easily cause excessively high pressures which pressure control valves are incapable of handling satisfactorily, resulting in expensive repair costs.

All types of hydraulic fluid increase in viscosity and eventually solidify as temperature decreases, though for various reasons:

(a) Naphthenic mineral oils and the majority of synthetic oils display a progressive viscosity increase without the appearance of a secondary solid phase as temperature falls, finally reaching a so-called 'viscosity pour point', i.e. the lowest temperature at which the oil still flows under standardized test conditions.

(b) Paraffinic mineral oils are subject to a gradual separation of small wax crystals as temperature decreases below the cloud point of the oil in question. In the absence of wax-modifying additives ('pour point depressants'), these crystals display a tendency to agglomerate and eventually form a rigid lattice, thereby preventing flow of the occluded oil phase. The lowest temperature at which the waxy oil still flows, again under standardized test conditions, is termed the 'wax pour point', and is to a certain degree dependent upon the previous time/temperature cycles or 'storage history'.

(c) Mineral oils contaminated by moisture may reveal an artificially elevated pour point due to formation of interlocking ice crystals as temperature falls below 0°C.

(d) Water-based fluids usually freeze with separation of two or more phases, ice crystals, additives, etc.

The principal volume of hydraulic fluids in use today is based on paraffinic mineral oils containing pour point depressant additives which modify the wax crystallization process, constraining any agglomeration tendency and preventing formation of an interlocking crystal lattice. Thus although numerous small wax crystals are apparent in the cooled oil phase, they are present in a form offering greatly reduced resistance to flow and the oil may be utilized at lower temperatures than the original base oil.

The very lowest natural pour points are inherent properties of certain naphthenic mineral oils and synthetic fluids ($\leqslant -60°C$), but it is important to appreciate that it is the actual *viscosity* at the lowest temperature of operation that constitutes the real criterion when selecting fluids for low temperature operation.

6.3 Temperature dependence of viscosity

The *viscosity index* (VI) expresses the gradient (slope) of the viscosity–temperature curve for a given fluid, and is of primary importance when considering hydraulic systems operating over a wide temperature range, for example units operating outdoors. The kinematic viscosity index (KVI, often abbreviated to VI) is calculated from the fluid viscosity measured in centistokes at 40°C and 100°C in accordance with ASTM test method D2270. This method is an extension of Dean and Davis' original procedure (*Chem. Met. Eng.* **36**, 318, 1928) in which the fluid was compared with viscosity data for reference oils of appreciably dissimilar viscosity–temperature gradients.

The temperature dependency of viscosity can be expressed by the Walther equation:

$$\log^2(v+a) = b \cdot \log T + c$$

in which v is the kinematic viscosity, $a \approx 0.8$, and b, c are constants for the fluid concerned. The Walther equation forms the basis for the standard viscosity diagrams, permitting viscosity–temperature data to be drawn as straight lines for Newtonian fluids. The graphical representation of viscosity–temperature relationships in this manner is obviously the most convenient basis for considering the viscosity variation of a fluid over a wide temperature span (Fig. 6.6).

The ideal hydraulic fluid would appear to be an equiviscous fluid, i.e. of constant viscosity independent of temperature variations. This is unfortunately an impossibility in practice, the temperature dependence of viscosity being a function of the actual chemical structure and eventual content of polymeric viscosity index improvers (VII) in the fluid.

Today, most hydraulic fluids are blended from highly refined mineral

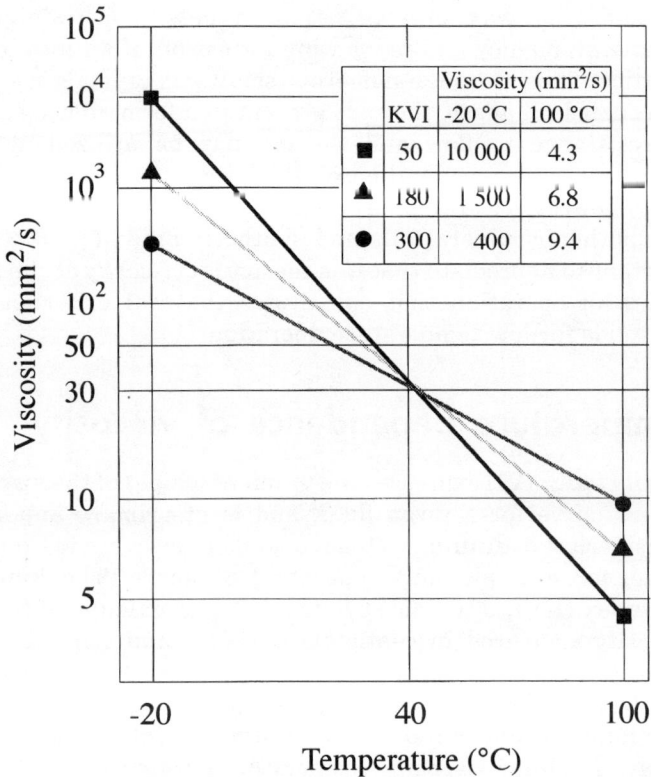

Fig. 6.6 Viscosity–temperature relationship of media with different viscosity indices.

base oils, with viscosity indices between 90 and 110. Certain synthetic oils, e.g. silicones, esters and polyolefins, also possess extremely favourable natural viscosity indices. By the addition of suitable viscosity index improvers to selected base fluids, products displaying viscosity indices around 350 may be prepared, permitting operation of suitable hydraulic equipment over a startlingly wide temperature range.

6.4 Shear stability

Rates of shear in hydraulic systems may be of the order of $10^7 s^{-1}$ in critical components, e.g. vane pumps, and this will have a pronounced effect on the viscosity of non-Newtonian fluids.

The effect of high shear rates is of particular interest in connection with the many polymer-modified mineral hydraulic oils in use, although other media, e.g. some silicones and water-in-oil emulsions, are also affected. Fluids containing high molecular weight polymers are subject to two viscosity-related effects as the shear rate increases. During the initial stage of streamlined flow, the large, predominantly linear polymer molecules show a progressive degree of orientation in the direction of fluid flow, this resulting in a reversible (temporary) viscosity loss. At more extreme high shear rates creating a turbulent flow pattern, the largest polymer molecules may also be mechanically degraded, forming smaller, less effective polymer fragments and resulting in an irreversible (permanent) viscosity loss (Fig. 4.3).

Research studies indicate that non-Newtonian fluids may be advantageous lubricants for plain bearings, maintaining the theoretical film thickness predicted for the low shear viscosity yet achieving significant ($\leqslant 40\%$) friction reduction. It is, however, essential to assess any potentially injurious effects of shear-related viscosity loss in critical components operating under boundary conditions, when considering use of polymer-modified hydraulic fluids of high viscosity index. High shear stresses occur in many locations in complex hydraulic systems, but it is often the pressure control valves that represent the most critical component in this respect.

Several factors influence the susceptibility of polymer additives towards high shear stresses, particularly molecular structure and molecular weight distribution. Commercial hydraulic fluids vary widely in their choice of base oil, polymer type and concentration, and hence in their performance under high shear conditions.

The ability of ISO HV (HR) type oils containing polymeric additives to withstand mechanical stresses and viscosity losses is normally reported in accordance with test method IP 294 or the corresponding DIN 51 382 (diesel injector test). For most practical hydraulic applications these test methods yield a relatively conservative indication of the viscosity loss to be expected in a real system (see Table 6.3).

These results clearly illustrate the importance of interpreting shear stability test data with caution when selecting fluids for exacting hydraulic systems. Commercial products normally show shear losses below 10% at 100°C when tested by the diesel injector method through 250 test cycles. In general the use of products limiting shear losses to a maximum of 5% is recommended for all hydraulic systems except non-critical, low pressure equipment.

Table 6.3 Comparison of shear stability by various test procedures (commercial ISO VG32 mineral oil with KVI≈180)

Test method	% Viscosity loss		Final viscosity (cSt at 100°C)
	At 40°C	At 100°C	
Diesel injector (IP 294): 250 cycles	3·4	4·1	6·5
FZG (IP 351) procedure C: 14h at load stage 3	13·0	14·7	5·9
Vane pump (IP 281): 250h at 14MPa	17	22	5·4

6.5 Pressure dependence of viscosity

The pressure dependence of viscosity is quite pronounced at large pressure changes. Figures 6.7 and 6.8 illustrate the viscosity of ISO VG46 and VG32 hydraulic oils (KVI=100) as a function of temperature and pressure. As depicted in Fig. 6.7, the viscosity of the oil at normal atmospheric pressure is approximately doubled when the pressure is increased to 35 MPa (350 bar).

The variation of viscosity with pressure is frequently neglected in low pressure hydraulic systems and where only moderate pressure variations occur. The effect is, however, quite appreciable at elevated operating pressures and the resultant viscosity increases may be of considerable significance with respect to filtration and lubrication. Furthermore, pressure surges due to hydraulic shock may attain values far greater than the normal operating pressure.

Chemical composition greatly influences the viscosity–pressure characteristics of a hydraulic fluid. In the case of mineral and synthetic oils the influence of pressure appears to correlate reasonably well with their respective viscosity indices. Hence the high viscosity index paraffinic oils are somewhat less affected by pressure than the low VI naphthenic oils of a predominantly cyclic molecular structure (Fig. 6.8). The least thickening tendency under pressure is shown by certain high VI synthetic oils, e.g. complex esters and polyethers.

The viscosity pressure coefficient (γ) is essentially constant for a wide range of fluids for the pressures between 0·1 and 70MPa of interest for hydraulics.

Numerous expressions have been suggested to predict the effects of temperature and pressure on the viscosities of various functional fluids. As may be expected, the more tractable the mathematical expression, the less accurate is the result.

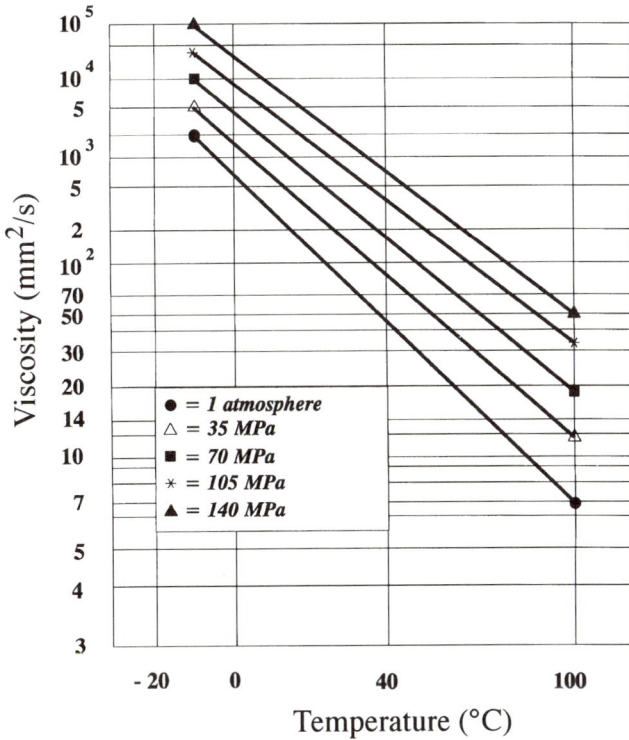

Fig. 6.7 Viscosity–pressure isobars of a typical mineral hydraulic fluid (Shell Tellus Oil 46).

One simple expression is

$$\eta = \eta_0 \exp{(\gamma p - \beta t)},$$

where η_0 is the viscosity at some reference temperature and pressure, p = pressure, t = temperature and γ and β are constants determined from measured viscosity data.

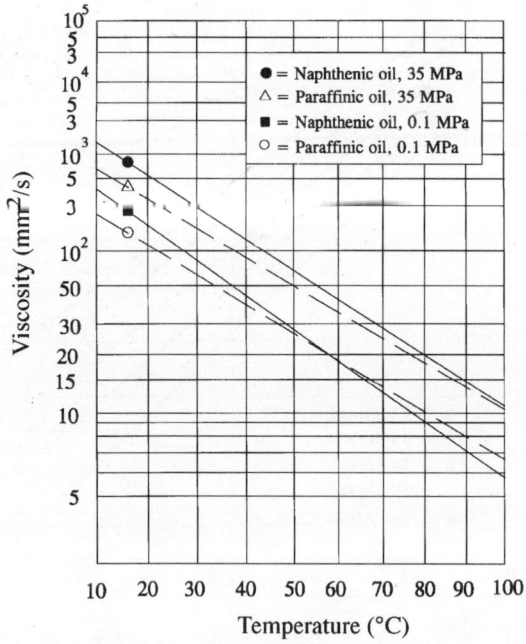

Fig. 6.8 Comparative effect of pressure on oils of low (naphthenic) and high (paraffinic) viscosity indices.

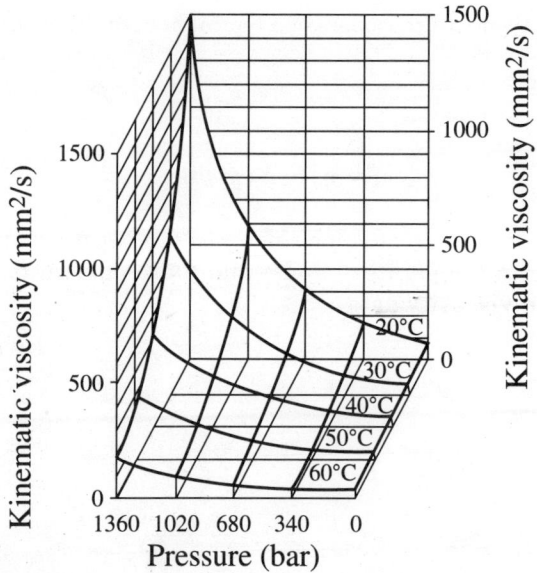

Fig. 6.9 Kinematic viscosity as a function of pressure and temperature for an ISO VG32 mineral oil.

7

COMPRESSIBILITY

The compressibility of a hydraulic fluid governs the energy absorbed by, and the time required for, generation of pressure by the system pump, and also the energy released on decompression. In addition, the compressibility of a hydraulic fluid is of prime importance regarding the *stiffness* of the power transmission and operation of servo systems. It is normally desirable that the compressibility should be as low as possible, although an extremely high resistance to compressibility in a hydraulic fluid could actually be disadvantageous in readily transmitting shock forces. However, the inherent elasticity of the mechanical components comprising the system have the effect of increasing the absolute compressibility of the fluid to a rather higher apparent value. In practice the requirement for an acceptably low

Table 7.1 Volume reduction of an ISO VG32 mineral oil (KVI=100) under pressure

Pressure (MPa)	Relative volume		
	10°C	40°C	100°C
0	1.000	1.000	1.000
25	0.989	0.986	0.981
50	0.978	0.974	0.966
75	0.969	0.964	0.953
100	0.961	0.955	0.943
125	0.953	0.947	0.933

compressibility is fulfilled by ensuring that the fluid is kept free from entrained air.

When a finite volume of liquid is compressed, its volume decreases by an amount (ΔV) proportional to the pressure increase (ΔP), i.e. a fractional volume reduction of $\Delta V/\Delta V_0$.

$$\text{Compressibility } (x) = -\Delta V/V_0 \cdot \Delta P$$

(the negative sign denotes a volume reduction). In other words the compressibility of a fluid is equivalent to the fractional volume reduction divided by the pressure change. For mineral oils this volume reduction is of the order of 0.5% for every 10 MPa pressure increase (Table 7.1).

The compressibility of mineral oils is slightly greater than that of water and varies, though only to a small extent, with oil type. Aromatic oils tend to be least compressible and naphthenic oils most, with paraffinic oils somewhere in between. Compressibility increases with rising temperature, while for oils of similar composition, those of lower viscosity are somewhat more compressible than the more viscous grades.

The reciprocal value of compressibility is termed the *bulk modulus* of the fluid:

$$\text{bulk modulus } (K) = -V_0 \cdot \Delta P/\Delta V$$

Since $V_0/\Delta V$ is a dimensionless fraction, the units of bulk modulus are the same as for ΔP, i.e. bar, N/m^2 or Pascal (Pa).

A high bulk modulus is synonymous with low compressibility, volume changes are relatively small and the fluid is stiffer in use.

The bulk modulus (K) of a fluid may be expressed as either a *secant* or *tangent* value (Figs 7.1 and 7.2) under *isothermal* or *isentropic* conditions. The isothermal or static bulk modulus, referring to compression without a change in temperature, is of significance only for relatively slow compression processes. (N.B. The notations B_T and B_S are sometimes used for bulk moduli under isothermal and isentropic conditions, respectively.)

7.1 Secant bulk modulus

7.1.1 The isothermal secant bulk modulus

The isothermal secant bulk modulus is the reciprocal value of compressibility referred to above, calculated for constant temperature conditions:

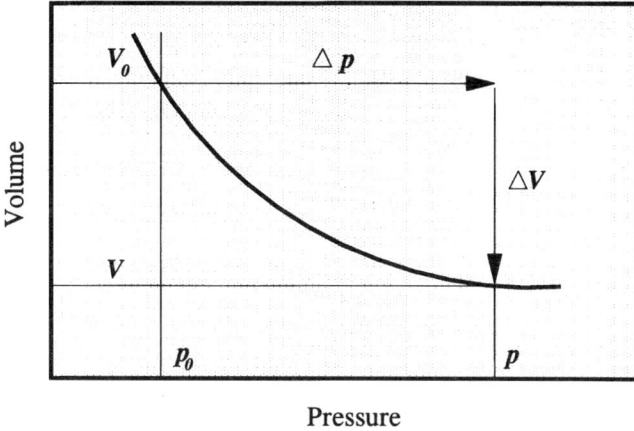

Fig. 7.1 Secant bulk modulus.

$$K_{\text{sec}} = -(V_0/\Delta V)\Delta P \text{ at constant temperature}$$

and yields an approximate value based on the total volume reduction over a given pressure change. This value can be calculated for any specific fluid from its density or viscosity at atmospheric pressure, as follows:

$$K_{\text{sec}} = \{1.30+0.15 \log\eta\}\{\text{antilog } 0.0023(20-t)\} \times 10^4 + 5.6P \text{ bar,}$$

$$K_{\text{sec}} = \{1.51+7(\rho-0.86)\}\{\text{antilog } 0.0023(20-t)\} \times 10^4 + 5.6P \text{ bar,}$$

where

P = pressure in bar,
t = temperature in °C,
η = kinematic viscosity at atmospheric pressure in cSt,
ρ = density at 20°C and atmospheric pressure in kg/l.

The *isentropic* (dynamic) *secant* bulk modulus is the preferred parameter where sudden variations in pressure and temperature of a fluid must be taken into account. Typical examples are calculating the forces acting on system components and reaction speeds of servo mechanisms:

$$K_{\text{sec}} = -(V_0/\Delta V)\Delta P \text{ at constant entropy}$$

As in the case of its isothermal counterpart, the isentropic bulk modulus may similarly be calculated by means of the kinematic viscosity or density of the fluid at atmospheric pressure:

$$K_{sec} = \{1.57 + 0.15 \log\eta\}\{antilog\ 0.0024(20-t)\} \times 10^4 + 5.6P \text{ bar}$$

$$K_{sec} = \{1.78 + 7(\rho - 0.86)\}\{antilog\ 0.0024(20-t)\} \times 10^4 + 5.6P \text{ bar}$$

7.2 Tangent bulk modulus

The *isothermal tangent* bulk modulus is the scientifically correct value at constant temperature. As the name suggests, the tangent modulus is derived from the gradient of the volume–pressure curve (Fig. 7.2) and is specific for the pressure P at which the gradient $-\mathrm{d}V/\mathrm{d}P$ is measured:

$$K_{tan} = -V(\mathrm{d}P/\mathrm{d}V) \text{ at constant temperature}$$

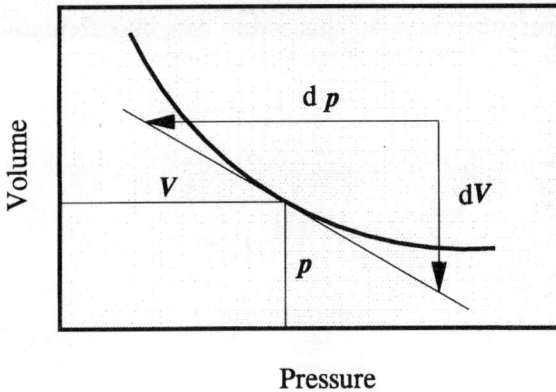

Fig. 7.2 Tangent bulk modulus.

The corresponding *isentropic tangent* bulk modulus may be calculated by multiplying the isothermal value by C_p/C_v, i.e. the ratio of the fluid's specific heat at constant pressure and volume, respectively. Typical values of C_p/C_v for mineral oils are quoted in Table 7.2.

Typical values of isentropic tangent bulk modulus for mineral oils at 50°C and atmospheric pressure are 1.62×10^3 MPa (ISO VG100) and 1.57×10^3 MPa (ISO VG32). Bulk modulus decreases with temperature, but increases with pressure (Fig. 7.3).

Table 7.2 Typical values of C_p/C_v for mineral oils

Temperature °C	C_p/C_v	
	Atmospheric pressure	70 MPa
10	1.175	1.15
60	1.166	1.14
120	1.155	1.13

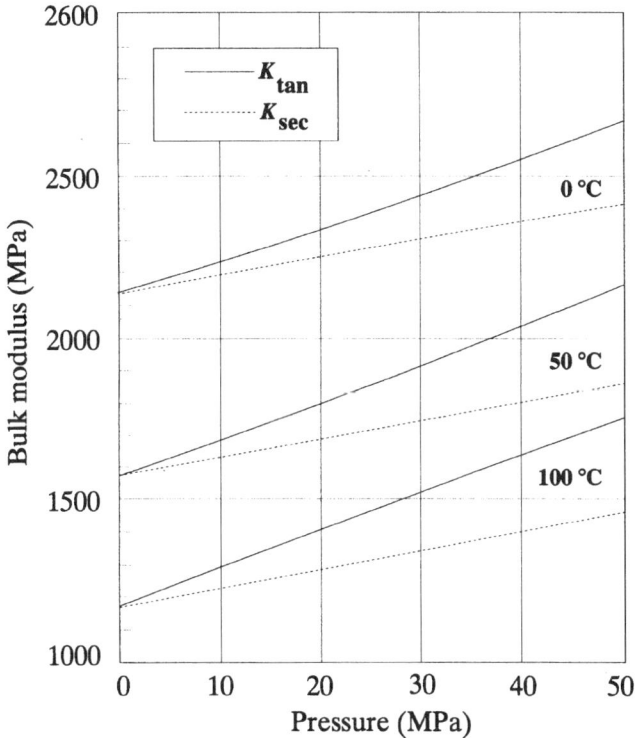

Fig. 7.3 Isentropic bulk moduli of ISO VG32 mineral oil, KVI≈100 (Shell Tellus Oil 32).

7.3 Effect of air on bulk modulus

At low pressures a content of free (undissolved) air in a hydraulic fluid results in dramatically increased compressibility and a correspondingly lower bulk modulus. This may be illustrated by the fact that the secant bulk modulus of a fluid containing 2% by volume of free air is reduced to approximately one-tenth of its original value at a pressure

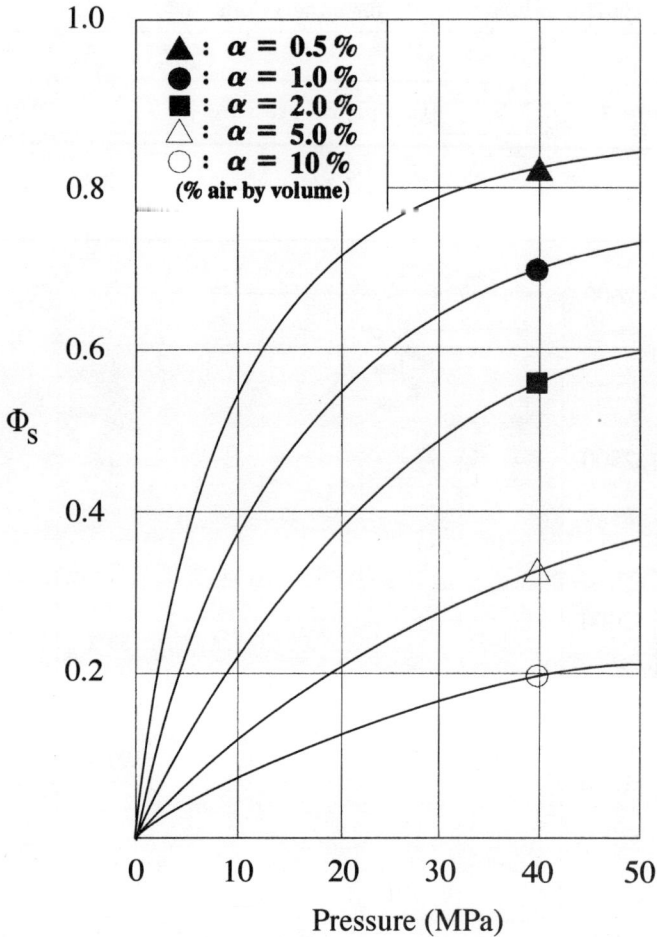

Fig. 7.4 Correction coefficient for secant bulk modulus of fluid containing undissolved air.

of 40 bar. As pressures increase, the effect of undissolved air decreases, and the bulk modulus is approximately halved by 1% undissolved air at a pressure of 170 bar. The factors influencing air entrainment and resulting deleterious effects, are reviewed in Chapter 10 (aeration problems).

Approximate values for the reduced bulk modulus resulting from a content of free air may be calculated by multiplying the normal air-free value by the respective coefficients Φ_{sec} and Φ_{tan} from Figs 7.4 and 7.5:

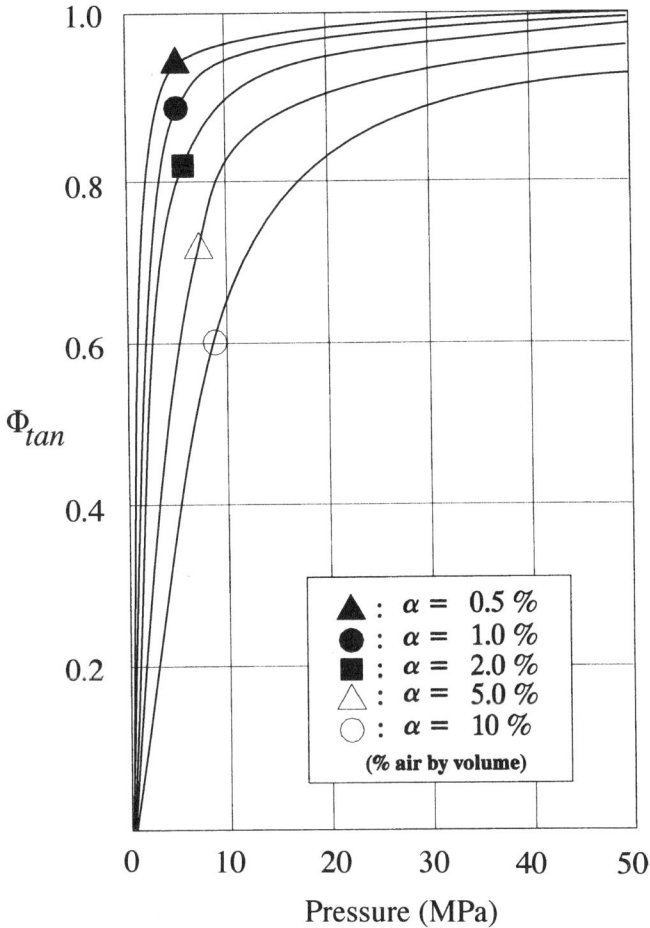

Fig. 7.5 Correction coefficient for tangent bulk modulus of fluid containing undissolved air.

$$\text{secant bulk modulus}_{\text{(aerated)}} = K_{\text{sec}}\Phi_{\text{sec}}$$

$$\text{tangent bulk modulus}_{\text{(aerated)}} = K_{\text{tan}}\Phi_{\text{tan}}$$

7.4 Low bulk moduli fluids

In contrast to the majority of hydraulic base fluids, a few fluids possessing generally favourable properties are characterized by surprisingly high compressibility. Silicone fluids, silicate esters, fluorocarbons and fluoroethers are examples of such fluids.

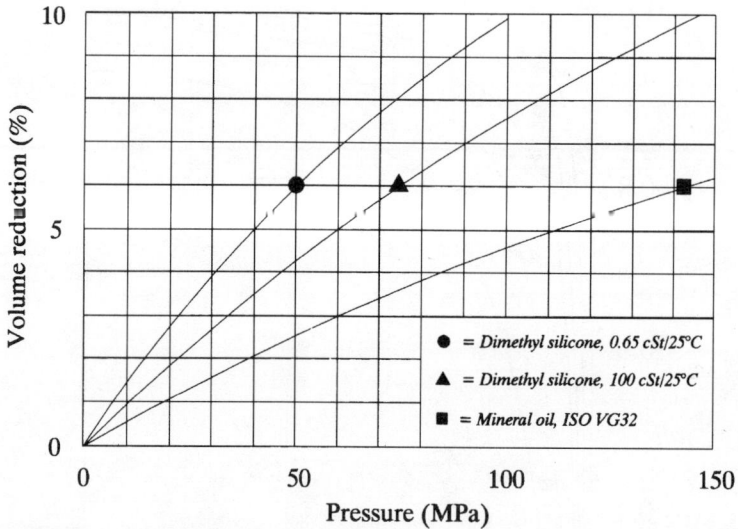

Fig. 7.6 High compressibility of silicone fluids.

Due to the flexible nature of the siloxane backbone, the silicone fluids exhibit a relatively high degree of compressibility and viscosity change under pressure. The low viscosity grades of the dimethyl polysiloxanes are consequently frequently utilized in shock-absorbing systems, and association with the helical molecular structure of the silicones has given rise to the pseudonym 'liquid spring'. Fig. 7.6 illustrates the high compressibility of silicone fluids compared to a conventional mineral oil.

7.5 Density

Density is denoted by the symbol ρ, and is the ratio of mass to volume at some specified temperature. The conventional unit is kg/l, and 15°C is a common reference temperature. All fluids increase their volume when heated and display a corresponding contraction on cooling. Due to this change in volume when subjected to temperature fluctuations, the density of the fluid varies correspondingly:

$$\rho_{t_2} = \rho_{t_1} - \alpha(t_2 - t_1),$$

where α is the density correction coefficient. Table 7.3 is an abbreviated version of Table 53 contained in *ASTM/IP Petroleum Measurement*

Table 7.3 Density correction coefficients for hydrocarbon fluids

Density at 15°C	Correction coefficent per 1°C (α)
0.7422–0.7534	0.000 79
0.7535–0.7646	0.000 77
0.7647–0.7757	0.000 76
0.7758–0.7866	0.000 74
0.7867–0.7984	0.000 72
0.7985–0.8020	0.000 70
0.8021–0.8279	0.000 68
0.8280–0.8594	0.000 67
0.8595–0.9245	0.000 65
0.9246–1.0243	0.000 63
1.0244–1.0742	0.000 61

Tables and is relevant for deriving values of η for mineral oil-based products and hydrocarbons. For high accuracy over wide temperature variations, the original ASTM/IP Table 53 should be used.

The variation in density with temperature (Fig. 7.7) has also been related to the average molecular weight of hydrocarbons and may be calculated with reasonable accuracy using the following equation:

$$\rho_{t_2} = \rho_{t_1} - \beta(t_2 - t_1) - \gamma(t_2 - t_1)^2,$$

where β and γ are functions of the molecular weight as given in Table 7.4.

The temperature-related variation of density has many practical applications in hydraulics, e.g. conversion of kinetic viscosities to dynamic ones, and efficiency calculations of kinetic transmissions.

The actual change in *volume* caused by temperature fluctuations may be calculated from the formula

$$\Delta V = V \cdot \alpha_{vol} \cdot \Delta t \text{ where } \Delta t \text{ is the temperature change}$$

The average volume expansion coefficient (α_{vol}) for mineral oil-based hydraulic fluids is 0.000 75 per °C. Therefore the volume of 1 litre of mineral hydraulic oil increases by 0.75 ml for a temperature rise of 1°C. This is of considerable significance with respect to the pressure of a fluid within a confined space, for example in a fully extended hydraulic cylinder. Under such conditions, with no possibility for the liquid to expand, the pressure will increase by approximately 1 MPa/°C. Consequently, even a moderate rise in temperature of 15°C would result in a volume increase of 1% and a pressure rise of 15 MPa (150 bar) in a confined space. If, for example, the heating effect

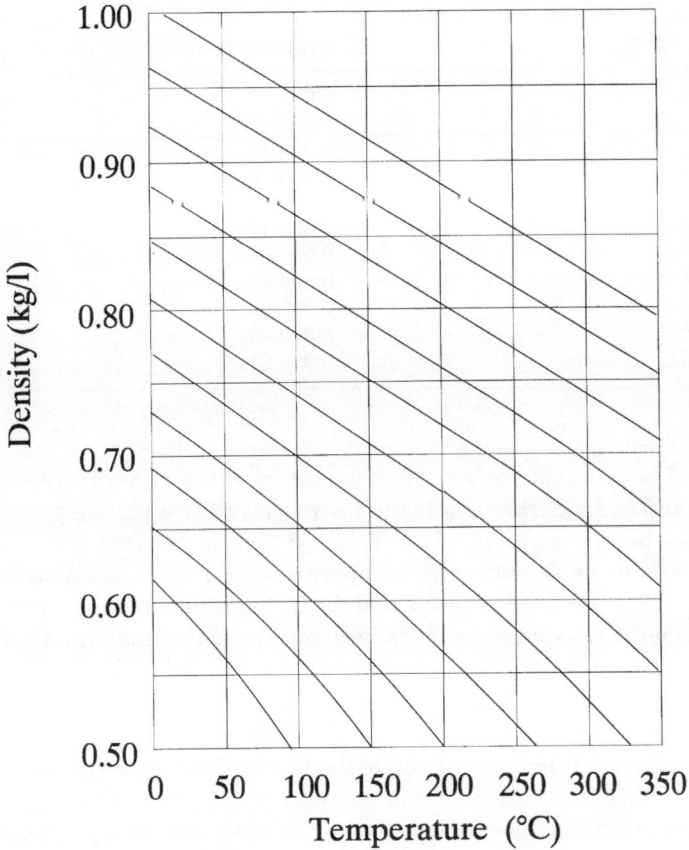

Fig. 7.7 Variation of density with temperature for mineral oils.

Table 7.4 MMW-related coefficients for calculation of density variation with temperature

Mean molecular weight (MMW)	β	γ
300	0.000 67	0.8×10^7
350	0.000 65	1.3×10^7
400	0.000 636	1.7×10^7

of the sun on an oil-filled hydraulic cylinder closed at both ends, causes a temperature rise of 40°C, the resultant pressure could increase to 40 MPa (400 bar)! For eventualities of this nature it is undoubtedly wise to consider safety measures to allow for pressure increases. Conversely, cooling of hydraulic fluids in a confined space will result in reduced pressures.

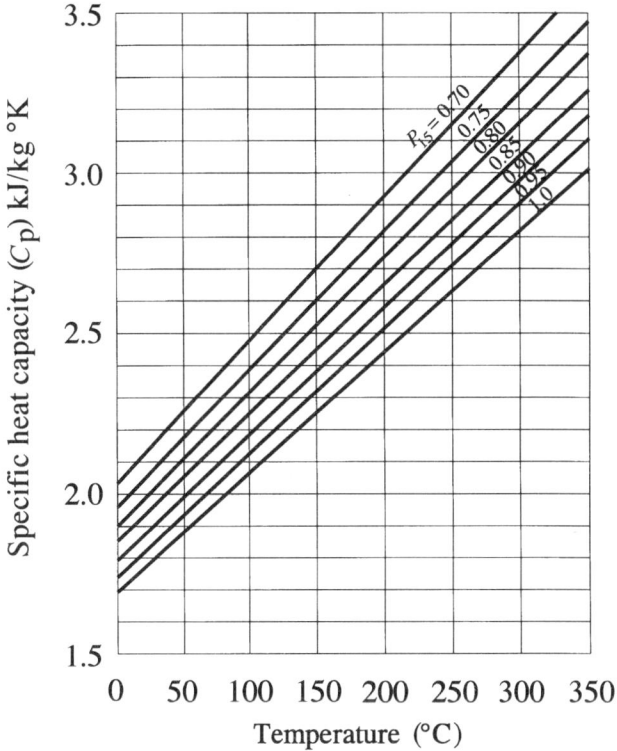

Figure 7.8 Specific heat capacity/density relationship of mineral oils.

7.6 Thermal properties

The *specific heat capacity* at constant pressure (C_p) is temperature-related, but insignificantly affected by the pressure level.

For mineral oils and synthetic hydrocarbons $C_p \approx 1841 + 4.4t$ J/kg per K (where t is in °C). Typical values of specific heat capacity for various fluid types are shown in Table 7.5.

Specific heat capacity is also function of density (Fig. 7.8).

The *thermal conductivity* (λ) is also temperature-related and is approximately 0.12–0.13 W/m per °K for mineral oils and synthetic hydrocarbons. Polarity and hydrogen bonding exert a significant influence on the properties of specific heat capacity and thermal conductivity; the respective properties for ISO type fluids HFA and HFC are thus significantly higher than for hydrocarbons (Table 7.5).

The relationship of thermal conductivity to density and temperature is illustrated in Fig. 7.9.

Table 7.5 Specific heat capacity (C_p) and thermal conductivity (λ) for various fluids

	Fire-resistant fluids				Synthetic oils	
	HFA	HFB	HFC	HFDR	Diester	Silicone
C_p(J/kg per °K)*	4000	3350	3300	1270	2310	1460
λ (W/m per °K)*	0.60	0.52	0.31	0.11	0.14	0.15

* Typical values.

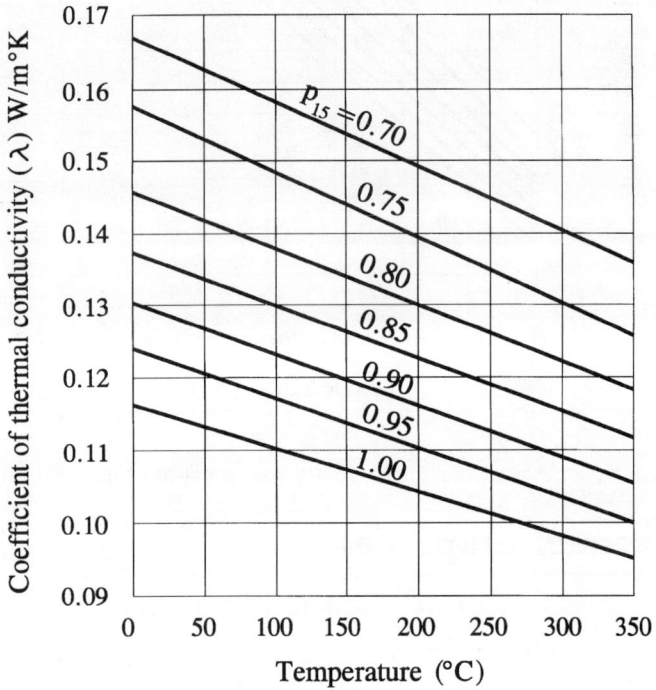

Figure 7.9 Thermal conductivity/density relationship of mineral oils.

8

ANTI-WEAR PROPERTIES

Even smoothly machined surfaces are quite rough, when examined by high power microscopy. As a result, mating surfaces do not achieve mutual contact over the entire surface, but touch only at the highest asperities. When a load is applied, these asperities are elastically, or even plastically deformed. The normal random distribution of asperity heights results in a progressive engagement of more asperities as the load increases. The area of true surface contact is thus proportional to the loading force. Under moderate loads, smooth sliding contact will tend to promote varying degrees of abrasive wear.

When mating surfaces slide over one another under such boundary conditions, the asperities in contact are not only deformed, but are frequently also heated to extremely high temperatures. These so-called local 'flash temperatures' may be so high that momentary welding of the lowest-melting asperity to its higher-melting counterpart occurs, a certain degree of metal transfer ensuing as continuing relative motion tears the weld apart, resulting in adhesive wear or 'scuffing'.

Despite the manufacturers' efforts to promote favourable conditions of operation within components, the many variables (pressures, temperatures, relative speeds and geometry of mating surfaces) frequently preclude hydrodynamic lubrication. Under the prevailing boundary conditions, frictional forces – and eventual wear – are largely dictated by the metallurgy of the contacting surfaces, lubricity of the hydraulic fluid and content of chemically active additives.

The addition of anti-wear additives to hydraulic fluids is principally to alleviate 'scuffing' wear between highly loaded sliding surfaces of steel on steel. Such additives are normally based on chemically bound

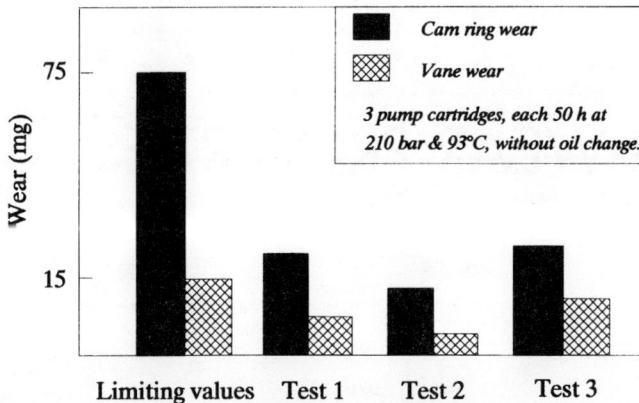

Fig. 8.1 Wear test, Vickers 35VQ25 (typical results).

phosphorus (aryl phosphates) or sulphur/phosphorus compounds that reduce mechanical wear by a form of 'chemical polishing'. This phenomenon involves the additive reacting with the metal surfaces wherever high local temperatures ('flash temperatures') occur due to friction between surface asperities in contact through the lubricating film. The resulting formation of a surface layer of reaction products prevents dynamic welding together of asperities and the subsequent catastrophic scuffing wear that would otherwise take place at highly loaded steel surfaces. This is of particular importance with respect to the stator ring and vanes of high pressure vane pumps and motors. Among the variety of standardized test methods to evaluate this aspect are the well-known tests utilizing Vickers pump types V104(5)-C and 35VQ25. In these methods the former pumps are usually operated at twice their normal rated pressure (140 bar). As system pressures have increased significantly during recent years, these tests are being superseded by the more demanding 35VQ25 procedure at 210 bar, which is considered more representative of modern equipment. Figure 8.1 shows typical test results for a fluid possessing satisfactory anti-wear properties.

Another test method, increasingly favoured within Europe, is the FZG (*Forschungsstelle für zahnräder und getriebebau, Munich*) gear test rig DIN 51 354, representing one of the main test criteria of HLP oil types as defined by the German specification DIN 51 524. This method evaluates the ability of the test fluid to prevent scuffing wear between spur gears under closely specified conditions of load, time and temperature (see Chapter 16).

Table 8.1 illustrates the progressive improvement of anti-wear properties resulting from the introduction of new additive types.

Table 8.1 Reduction of wear by use of improved additive systems

Additive system	With rust and oxidation inhibitors	With arylphosphate	With + ZDTP
Vane pump test IP 281 (250h, 17Mpa, 1450rpm) Total wear (mg):	690	240	30
FZG test – IP 334 (Type A/8.3/90°C) Failure load stage:	6	8	10–12
Shell 4-ball test – IP 239 Wear scar, 20kg load (mm):	0.60	0.30	0.30
–/– –/–, 40kg load (mm):	Welding	Welding	0.39

In addition to the test methods mentioned above, many other standardized and in-house procedures are specified by component manufacturers for evaluating the anti-wear properties of hydraulic media.

It is imperative to exercise caution when selecting anti-wear additives for a specified application, in order to avoid undesirable side effects, for example less satisfactory frictional characteristics with certain other metal combinations or reduced corrosion inhibition. Sliding contacts of yellow metals–steel appear more sensitive towards the viscosity of the lubricant than its composition, though under high load conditions the more chemically active additives are notably less satisfactory. One classic problem area in this respect is the disproportionately high wear of phosphor bronze slipper pads in certain high efficiency piston pumps. Research in latter years has been directed towards developing additive systems affording satisfactory protection to the more sensitive metals, whilst at the same time achieving adequate performance in the important FZG test. This desirable combination of properties is commonly referred to as 'multi-metal compatibility'.

The successive evolution of anti-wear additive systems through the years, is depicted in Fig. 8.2.

Most yellow metals (copper, bronzes, brass) are best lubricated by fluids containing relatively mild anti-wear additives, or even by turbine oils without any form of chemically active additive.

Aluminium and alu-alloys seem to be generally compatible with most of the usual additive systems utilized in hydraulic fluids (but not with all base fluids), although highly loaded combinations of aluminium bronze–steel may be problematic. Silver overlays on bearing surfaces,

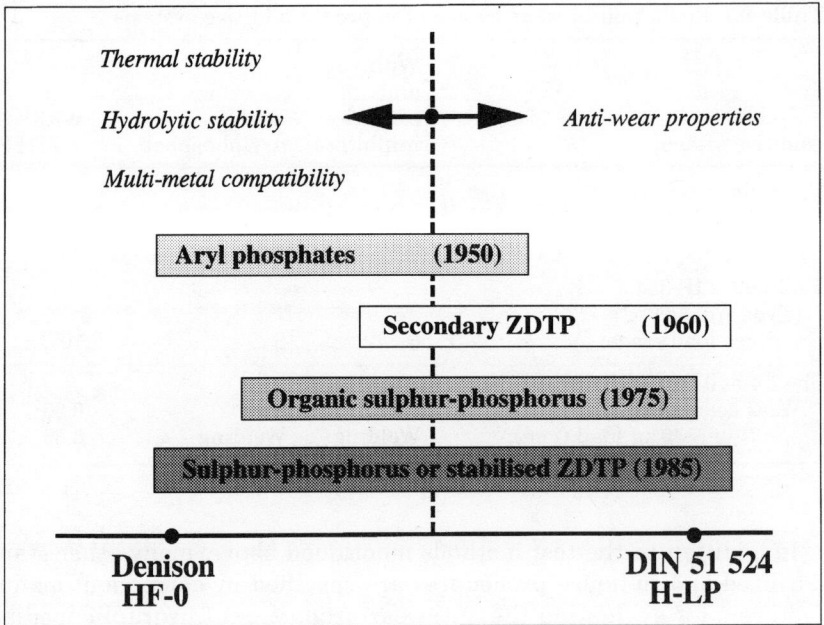

Fig. 8.2 Development of anti-wear additive systems.

less common today, are rapidly attacked by the active dithiopho-sphate additives, and fluids formulated using relatively mild addi-tives, e.g. ISO type HL or turbine oils, may be preferable when silver is present.

9

OXIDATION STABILITY

An extremely important criterion concerning the quality of hydraulic fluids, particularly in high temperature applications, is the ability of the fluid to resist chemical degradation by reaction with atmospheric oxygen. The oxidation stability of a hydraulic fluid thus determines the resistance to ageing, i.e. the useful 'life' of the fluid.

Oxidation is a notably exothermic process, and in the case of mineral oils, extremely complex in nature due to the prolific variety of molecular structures involved. The degradation of hydraulic fluids by oxidation can result in significant viscosity increases, development of corrosive organic acids, and lacquering of critical valve surfaces, etc., by resinous oxidation products.

Oxidation of mineral oils is a progressive chemical process in which hydrocarbon molecules are attacked by oxygen and broken down to a multitude of various oxygenated compounds, including hydroperoxides, alcohols and organic acids. The oxidation process is principally influenced by fluid composition, temperature, oxygen (air) availability and catalyst metals such as copper and iron.

Oxidative degradation is not normally a major problem with the better-quality hydraulic fluids of today. Hydraulic oils are usually based on highly refined base oils, and even though the oxidation rate is an exponential function of temperature (the velocity doubles for every 8–9°C rise in temperature), the operational temperature of most systems will seldom exceed 65°C. At this temperature, the oxidation-limited life of an inhibited hydraulic oil of good quality would be of the order of several thousand hours (Fig. 9.1).

Hydraulic media designed for high operating temperatures require

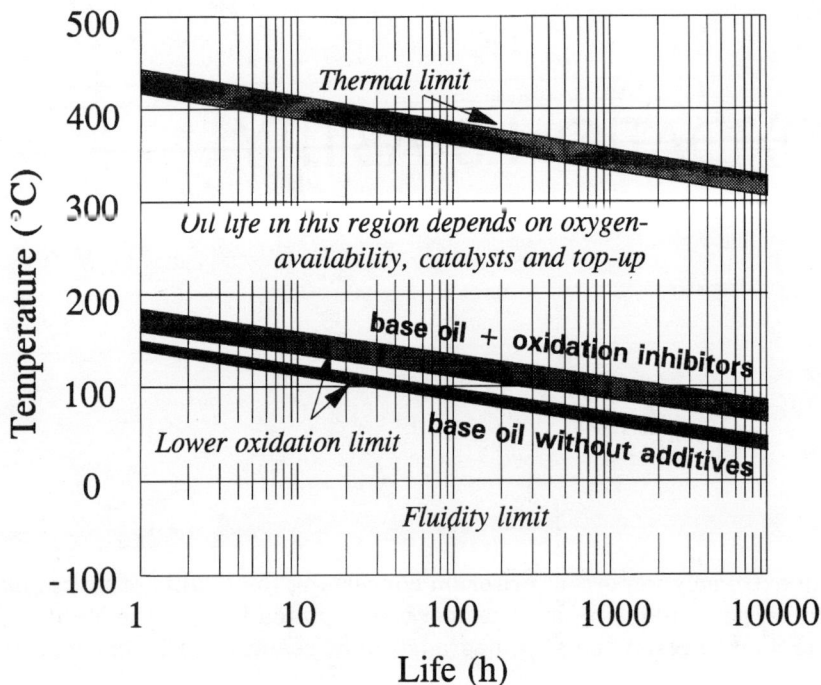

Fig. 9.1 Temperature limits for mineral oils.

careful selection of both base fluid and additives. High quality mineral base oils, suitably inhibited, yield satisfactory performance at operating temperatures up to ≈90°C, although in the upper temperature range it becomes increasingly important not to prolong replacement of the fluid unduly, especially in systems incorporating sensitive valve assemblies. For extended operation at temperatures above 90°C, use of selected synthetic base fluids, e.g. polyalphaolefins, confers distinct advantages with respect to oxidation stability and volatility. Oxidation inhibitors such as sterically hindered phenols, alkylated diphenylamines, metal passivating sulphur/phosphorus compounds, etc., can extend the induction time (Fig. 9.2) considerably and thereby delay the formation of injurious oxidation products.

Oxidation of mineral oils is an unusually complex process, not least because a typical mineral base oil is composed of several thousand isomeric hydrocarbon homologues. Chemically, thermal oxidation of hydrocarbons is an autocatalytic reaction, initiated by peroxide or hydroperoxide formation, proceeding as a chain reaction

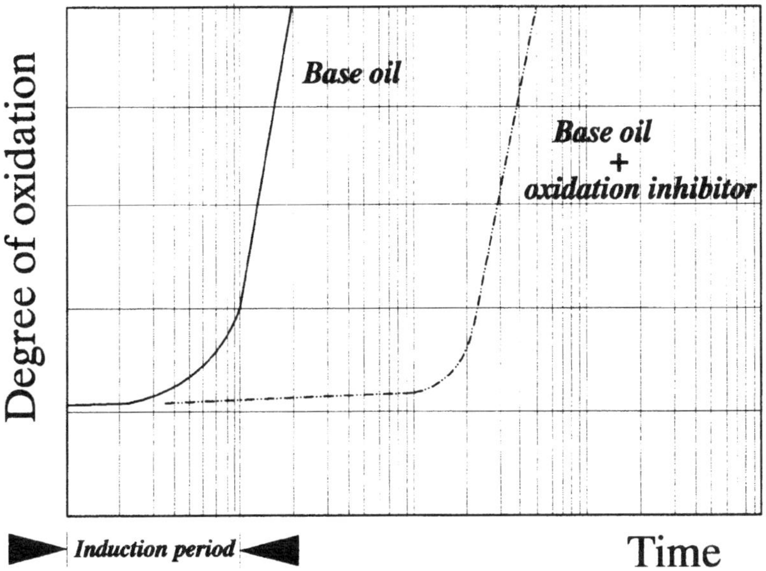

Fig. 9.2 Oxidation curves for oils with/without oxidation inhibitor.

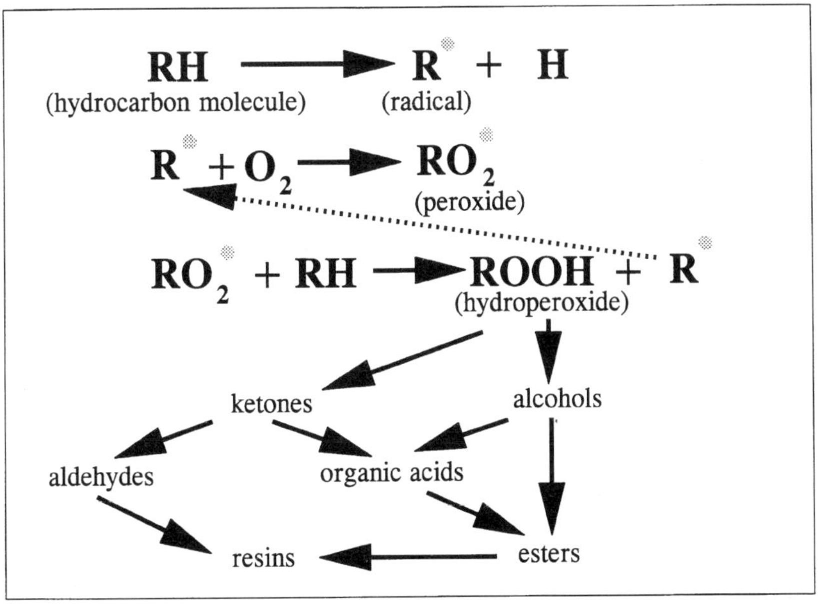

Fig. 9.3 Oxidation of mineral oil. ✳ = Energy-rich radical or molecule.

and developing an increasing complexity of branches as intermediate oxidation products react together.

Figure 9.3 illustrates how the peroxide compound, formed at an early stage of the process from a dissociated hydrocarbon molecule, attacks at random another hydrocarbon molecule, simultaneously liberating a new energy-rich free radical which rapidly oxidizes to produce a fresh, reactive peroxide.

Intermediate reaction products of hydrocarbon oxidation, such as alcohols and ketones, are subject to further degradation, forming carboxylic acids, keto- and hydroxy-carboxylic acids. Other intermediates, e.g. aldehydes, eventually form resinous deposits by complex condensation reactions.

The intricate oxidation process is further complicated by the accelerating effect of catalyst metals. Particularly dangerous catalyst materials are soluble metal soaps, which may gradually be produced from copper, lead, zinc and ferrous metals in the system by corrosive oxidation products. Many hydraulic fluids incorporate special additives to counteract the catalytic effect of metals (metal passivators) and soluble soaps (chelating additives) (see Chapter 4).

From the above remarks it will be readily appreciated that the oxidation reaction sequence and associated kinetics can only be summarized in very broad terms. Mineral base oils of high viscosity index, i.e. composed predominantly of paraffinic hydrocarbons, tend to produce a higher proportion of acids than insoluble resinous materials. Conversely, oxidation of naphthenic base oils with a higher aromatic content results in proportionately larger amounts of insoluble 'asphaltenes'. Of direct significance for hydraulic systems is the fact that agressive organic acids and insoluble resinous compounds are the final products, and also the accelerating effect of higher temperatures and catalyst materials. Figure 9.4 illustrates the variation of oxidation product distribution with time for a synthetic base oil, oxidized at 110°C.

Various laboratory test procedures are used to evaluate the oxidation stability of hydraulic fluids, and predict their oxidation-limited useful life (see Chapter 16). Selection of suitable methods and correct interpretation of test results, in relation to product formulations and applications, require a certain degree of expertise. Examples of this are the greatly misleading results obtained when applying the well-known TOST test (ASTM D943) to certain biodegradable hydraulic fluids, and indiscriminate use of IR-spectroscopy to assess additive depletion. In the first instance, fatty oils hydrolyse, producing unrealistically high acidities under the conditions of the TOST procedure, whilst in the latter example certain additives continue to exert their

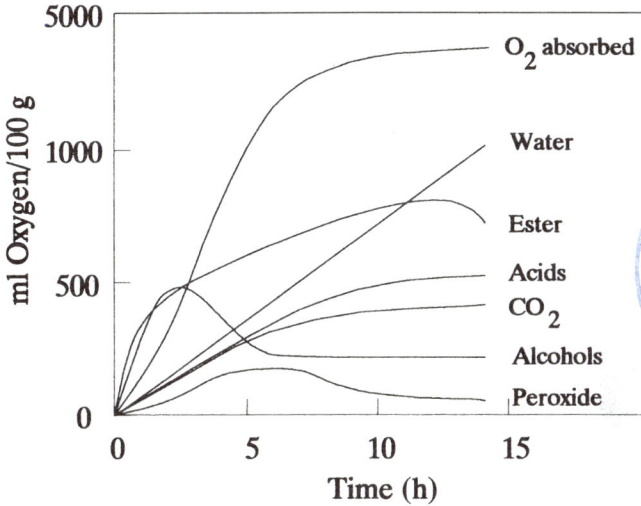

Fig. 9.4 Variation of oxidation products with time (polyisobutene).

inhibitive action via intermediate reaction products, despite a significantly reduced concentration of the initial additive.

The type of refining process has a pronounced influence on the subsequent behaviour of a base oil during oxidation. This is illustrated in Fig. 9.5, which compares the oxidative deterioration of (a) a severely hydroprocessed mineral oil, and (b) a solvent-extracted base oil from the same crude oil, containing the same oxidation inhibitor. The severe hydroprocessing of oil results in almost total removal of all sulphur compounds, including natural oxidation inhibitors. This secondary effect is unfortunate as many sulphur compounds, e.g. thiophenes and organic sulphides, are extremely beneficial, imparting valuable oxidation-inhibiting properties to the refined base oil. Conversely, solvent extraction effects a far more selective removal of undesirable components in oil, a suitable concentration of natural inhibitors remaining in the refined oil.

As the figure shows, hydroprocessed oil responds satisfactorily to the added oxidation inhibitor, but oxidative deterioration progresses rapidly as soon as the induction period is past. Although the solvent-extracted grade does not display quite as good an initial resistance to oxidation, the content of natural sulphur compounds continues to inhibit the oxidation rate even though the added oxidation inhibitor is no longer functional. Thus the oxidation rate increases more gradually than is the case with the hydroprocessed product, and provides more time to detect the final stages of the fluid's useful service life.

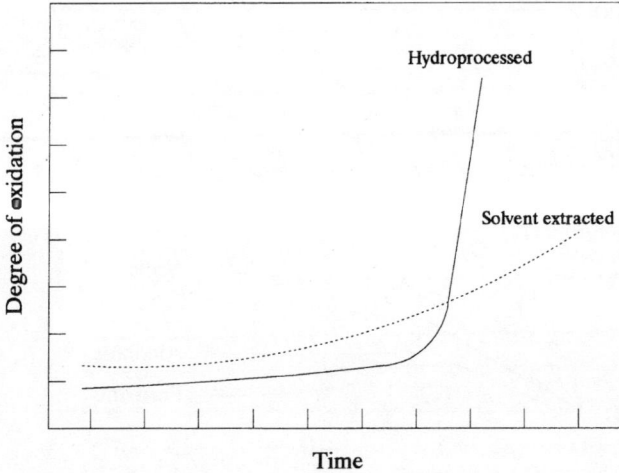

Fig. 9.5 Influence of refining process on oxidation characteristics of mineral base oils.

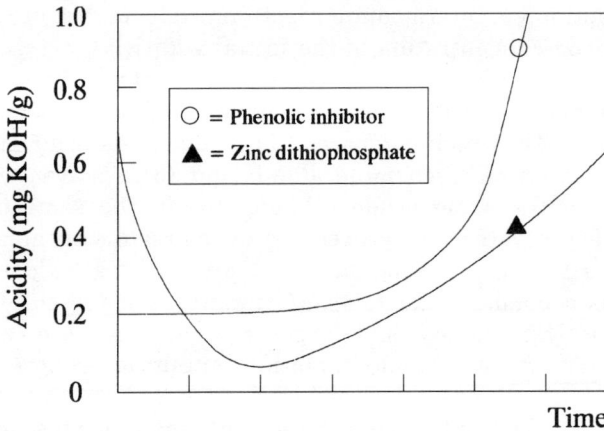

Fig. 9.6 Oxidation acidity curves for different types of oxidation inhibitor.

During routine condition monitoring of used hydraulic oils, it is usual to determine the degree of oxidation by measuring increases in viscosity, acid value and content of high molecular weight resins. The initial acidity of a fluid, and its subsequent development with time as the oil ages, is greatly dependent upon the additives present. Oils formulated with the two most familiar oxidation inhibitor types, i.e. hindered phenols and metal dithiophosphates, display greatly differing acidity–time relationships (Fig. 9.6).

When utilizing phenolic oxidation inhibitors, the initial acidity is principally derived from other components, e.g. rust inhibitors. The neutralization number of the fluid remains fairly constant at a relatively low value until the induction period is passed, whereupon the acidity rises at an increasing rate.

Dithiophosphate inhibitors, however, impart a natural high initial acidity to the fluid, normally between 0.6 and 1.5 mg KOH/g in hydraulic fluid formulations. This initial high acidity stems from the amphoteric nature of metal dithiophosphates and decreases to a minimum value during the initial period of operation as the additive decomposes and removes active peroxide radicals from the fluid. Concurrent with this function, thermal degradation of the multi-functional dithiophosphates at rubbing surfaces results in the formation of surface films with effective anti-wear and metal passivating properties. When the content of oxidation inhibitor and active decomposition products in the fluid are finally depleted, the oxidation reactions gain momentum, resulting in rising acidity as illustrated in Fig. 9.6.

10

DEMULSIBILITY

Apart from special applications using water-based fluids, it is normally desirable to exclude water as far as possible from hydraulic systems. In practice this is not an easy task, particularly where open-vented systems are concerned, i.e. in the majority of industrial installations. Water may gain access to the hydraulic fluid by a number of routes, including condensation from humid air drawn in via breather vents, leakage from oil-coolers, and inadvertant filling of moisture-contaminated hydraulic fluid from drums or containers stored incorrectly. Significant accumulations of condensation water have been registered in hydraulic fluid reservoirs where the upper surfaces have been exposed to cooling air from nearby ventilation fans, and also in systems where cooling coils were partly exposed to the free air volume above the fluid level. The ability of a typical mineral hydraulic oil to absorb moisture is illustrated in Fig. 10.1.

There are several possible effects of moisture contamination, the most obvious being the formation of rust and subsequent mechanical problems. In addition to the destructive effect of surface wear, rust particles can promote the formation of oxidation sludge and deposits which cause serious malfunctioning of servo-valves, pumps, motors and other critical components. Rust particles circulating with the hydraulic fluid may cause abrasive wear and accentuate any tendency towards cavitation erosion at valves, pipe elbows and other locations where high fluid velocities and directional changes are experienced. Formation or accumulation of rust against dynamic seals results in accelerated wear of the seals and increasing leakage of the hydraulic fluid. On account of the tendency of dispersed water to encourage

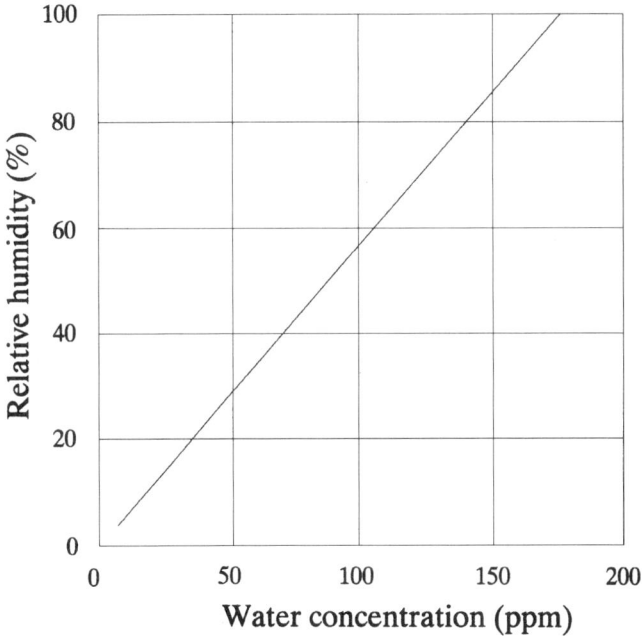

Fig. 10.1 Equilibrum water content of a typical ISO VG32 mineral oil at 40°C as a function of humidity.

circulation of mechanical impurities within the system, and initiate hydrolysis reactions with additives, poor demulsibility may indirectly promote filter blocking.

The deleterious effect of both free and dissolved water on the fatigue strength of ball bearings and similarly stressed alloy steel components is well documented, proceeding by a hydrogen embrittlement mechanism. The effect may be quite significant, tests confirming that as little as 0.01% (100 ppm) of dissolved water can reduce the fatigue life by 32–48%. The vast majority of industrial hydraulic systems operate at higher water contents, some free water is normally present and the concentration of dissolved water in the fluid is at saturation level. Thus the potential danger of decreased fatigue life may be very relevant with respect to hydraulic pumps and motors incorporating heavily stressed ball and roller bearings or gears.

In order to alleviate the deleterious effects of water contamination, all possible means should be exploited to minimize the moisture content of the system. An invaluable design feature is the positioning of a drain cock at the lowest point of the fluid reservoir, whereby undissolved water (a certain amount will be precipitated out of solution as the temperature falls after shut-down), and sludge and extraneous material may be periodically drained from the system. Unfortunately, the floor of fluid reservoirs rarely slopes evenly towards the drain cock; small deformations, inadequate levelling during installation and the presence of solid baffle plates within the reservoir can all reduce the efficiency of removing water and bottom sediment by drainage.

For systems subject only to incidental water contamination, e.g. by condensation, it is generally accepted that a hydraulic fluid should display good demulsibility, separating rapidly from entrained water during stationary periods in the fluid reservoir. In general, best demulsibility is achieved by the most highly refined mineral base oils and certain synthetic oils, e.g. polysiloxanes and fluorocarbons, with negligible affinity towards water. Most additives detract from the hydrophobic character of the pure base fluids to some degree, and display an adverse effect on demulsibility. As many of the more effective rust inhibitors are polar compounds of a somewhat hydrophilic nature, the desirable requirements for good demulsibility together with excellent protection against rust formation are often difficult to harmonize. The initial ability of a fluid to separate from dispersed moisture gradually deteriorates in use as the concentration of polar oxidation products increases.

The most troublesome hindrance to good demulsibility is nevertheless the presence of contaminants in the form of soluble emulsifiers and finely divided particulate impurities. Emulsifiers are essentially chemical substances possessing an affinity for both the fluid and the aqueous phase. With respect to the hydrophobic fluids mentioned above, the molecular structure of an effective emulsifier is characterized by a non-polar, hydrocarbon concentration of atoms in one part of the molecule, and hydrophilic groups (e.g. -carboxyl, -amine, or -hydroxyl) groups in a different part of the emulsifier molecule. These substances therefore have a simultaneous affinity to both water and hydrocarbon fluid phases, the efficiency of the emulsifier depending largely upon its ability to reduce the interfacial tension between the two phases, thereby increasing their mutual adhesion. Stability of such a water-in-oil emulsion increases as the adsorbed layer of emulsifier molecules around the dispersed water droplets becomes more tightly packed, and thus less prone to displacement from the interface. Regrettably, little can be done to remedy poor demulsibility caused by soluble emulsifiers. Addition of 'emulsion breakers' is a theoretical possibility, though practical experience indicates that the chances of selecting a suitable dosage of a suitable additive, and achieving long-term improvement, are slender.

Practical experience with hydraulic systems indicates that in many instances deteriorating demulsibility is caused by the formation of metal soaps during use, predominantly calcium and zinc compounds.

Zinc is, of course, used extensively as an anti-wear agent and oxidation inhibitor in hydraulic fluids. Injudicious topping-up with alternative products containing, for example, acidic rust inhibitors, can in the presence of moisture result in the formation of zinc soaps, with adverse effect on demulsibility. Mixing fluids of dissimilar composition should therefore be avoided unless approved by the supplier. Use of galvanized plating, pipes and components is not recommended, due to the risk of soap formation by reaction with organic acids. Serious emulsification (and air entrainment) problems may also be experienced if dispersant type engine oils are inadvertently added to a hydraulic system.

The presence of small amounts of colloidal particles, e.g. chalk or rust, may also significantly reduce demulsification due to stabilization of the interface between the aqueous and fluid phases. A typical example of this phenomenon occurred in a plastics factory, where the hydraulic oil was found to be contaminated with microscopic particles of the plastic moulding powders, drawn into the fluid reservoir via a breather vent. The demulsibility of a fluid contaminated by particles can often be significantly improved by filtration.

Some systems are also subject to a relatively significant contamination by water, e.g. those operating in extremely humid atmospheres, in paper mills and where leakages of cooling water continually occur. In such cases a hydraulic fluid formulated to provide a limited degree of detergency may be preferable. Hydraulic fluids of this nature are, for example, classified as type HLP-D in the German specifications. These detergent hydraulic fluids effectively prevent harmful accumulations of water in the hydraulic circuits and components, yet by virtue of the controlled detergency permit sedimentation of dispersed water in the fluid reservoir from where it may be drained off at frequent intervals.

The major filter manufacturers also supply special filter elements capable of removing limited amounts of free water from hydraulic fluids. Typical designs consist of a water-absorbing polymer enveloped by a porous membrane of polypropylene. The polymer combines chemically with free water without softening or leaching into the hydraulic fluid. Such filters can be an effective means of controlling the moisture content of an average system, but are hardly an economic solution for systems subject to continuous contamination by significant quantities of water.

10.1 Protection against corrosion

Modern hydraulic systems bear little resemblance to the relatively simple units of some 50 years ago, incorporating complicated valves and other components which may often be particularly sensitive to malfunctioning induced by corrosion or fine particulate contaminants derived from corrosion. Consequently, it is necessary to supplement the hydraulic fluid's inherent corrosion protective properties by addition of an effective corrosion inhibitor. In mineral oil products, the inhibitors are often polar compounds, for instance, long chain dicarboxylic acids and their salts, forming strongly adsorbed multi-molecular films on all oil-wetted surfaces. Assuming an adequate concentration, the combination of steric configuration, molecular dimensions and polarity of the inhibitor presents an impermeable barrier to corrosive substances, moisture, etc. The protective mechanism may be compared to a host of tadpoles swimming to the sides of a bucket, resting their heads against the confining surface and waving their tails, thus preventing further access to the surface. Many polar additives display a pronounced deleterious effect on the demulsifying properties of hydraulic fluids, and certain acidic corrosion inhibitors attack galvanized surfaces forming zinc soaps. As in the case of other additives, it is therefore prudent to select corrosion inhibitors with care.

Corrosion inhibitors utilized in water-based media are often oxidizing salts which reduce the electronegative character of anodic surfaces, pH-regulators (buffers) and various polar additives forming adsorbed surface films.

11

AERATION PROBLEMS

Solubility of air in a liquid is directly proportional to the absolute pressure above the liquid surface (Henry's law), and normally decreases with rising temperature. The solubility is often quoted in the form of a 'Bunsen coefficient' (α), equivalent to parts by volume of gas in the liquid at 0°C and normal atmospheric pressure, i.e. 760 mm Hg.

As shown in Fig. 11.1, the solubility of air in mineral oil is relatively high compared to other typical hydraulic media. Despite its high solubility, air is of minor significance as long as it remains dissolved in the hydraulic fluid (the Bunsen coefficient is as high as 0.08–0.09 for mineral oils, but only 0.02 for water). Pressure changes in the system may, however, lead to formation of free air bubbles, these resulting in a dramatic increase in compressibility of the fluid at low operating pressures. For instance, at a pressure of 30 bar the secant bulk modulus would be reduced by approximately 80% for a mineral oil containing 1% free air. The effect of aeration on compressibility decreases with higher operating pressures as the specific volume of undissolved air remaining in the oil progressively decreases. Thus at an operating pressure of 230 bar the corresponding bulk modulus would be reduced to approximately 60% of the original air-free value.

Undissolved air in a hydraulic fluid will of course reduce the density of the fluid by an amount equivalent to the volume fraction of air present. Finely dispersed air bubbles have also been found to increase the viscosity of a hydraulic fluid slightly, though this effect is without practical significance.

The practical consequences of excessive aeration in a hydraulic

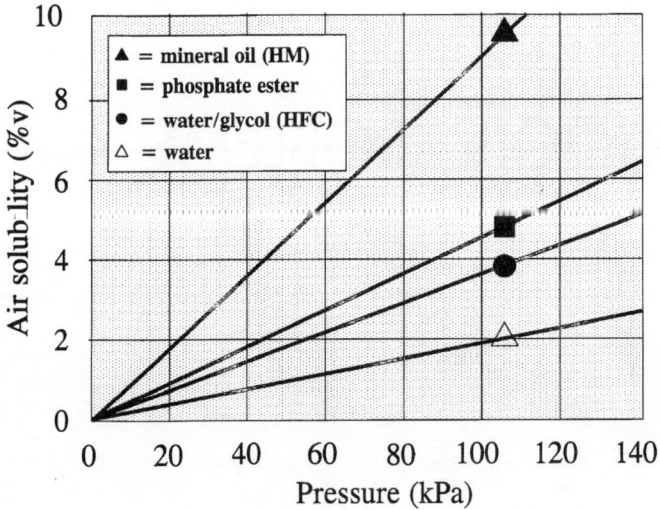

Fig. 11.1 Solubility of air in various media.

system become painfully obvious in the form of high noise levels, unreliable operation, and possible damage to pumps and other components.

Undissolved air is a frequent cause of cavitation on the suction side of hydraulic pumps, and collapsing air bubbles on the pressure side may often give rise to damage resembling that associated with ordinary cavitation erosion. These phenomena occur at, or in the near vicinity of, pressure control valves and bearings lubricated by the hydraulic fluid. Undissolved air will normally also result in a certain pressure drop and a tendency for laminar flow to be replaced by turbulent conditions.

Free air bubbles passing through a hydraulic pump are subjected to sudden compression under adiabatic conditions, thus raising the temperature of the compressed bubbles by several hundred degrees centigrade (see Fig. 11.2). Consequently, the film of liquid surrounding the hot bubble is prone to severe thermal degradation ('cracking') and oxidation. In the most serious cases hydrocarbon-based fluids assume a grey–black colour, significant viscosity reductions occur and the oil has a distinctly burnt odour. Examination of the oil under a microscope will frequently reveal a multitude of near-spherical carbon 'skeletons' resulting from the thermal cracking.

Frequent causes of foaming and circulation of undissolved air are

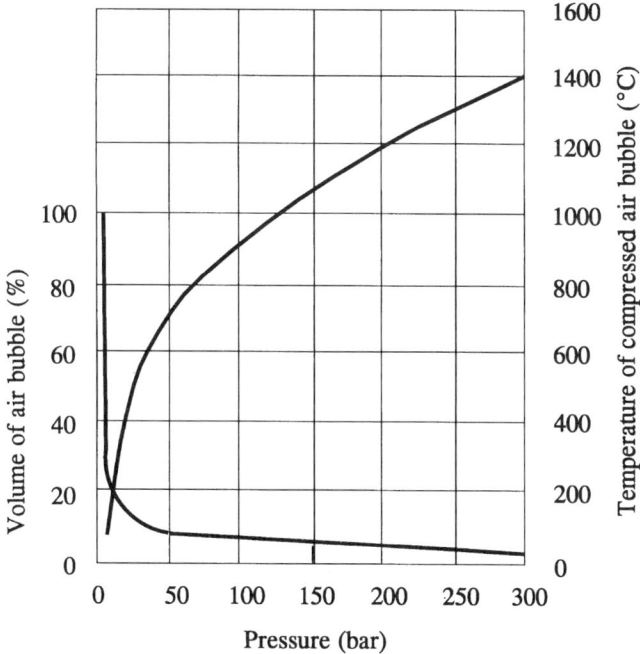

Fig. 11.2 Temperature of dispersed air following adiabatic compression.

- leakages in suction lines, pipe connections, glands, etc.,
- return fluid (which often already contains free air) splashing freely down into the fluid reservoir,
- return fluid which is permitted to flow unhindered to the pump intake,
- low fluid level in reservoir, insufficient residence time,
- contaminants, finely divided particles and dissolved impurities.

As is the case with many other types of contaminant, finely divided particles can display a stabilizing effect on air bubbles (and free water) in the fluid.

Adverse effects may also result from contamination by other fluids containing additives which reduce the ability of the hydraulic medium to release undissolved air. For example, topping up with a *detergent engine oil* will often have an extremely undesirable effect on the air-release (and demulsifying) properties of a premium hydraulic oil.

The ability of a hydraulic system to alleviate aeration problems is also largely dependent upon design and maintenance procedures.

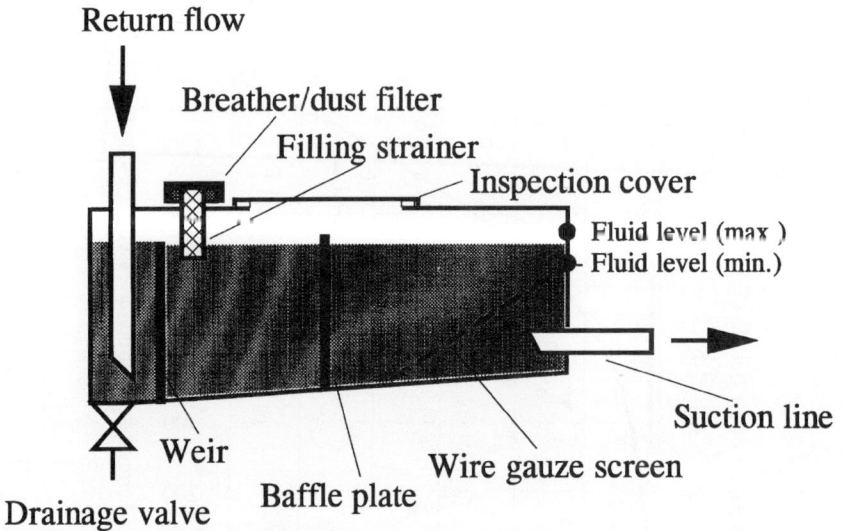

Fig. 11.3 Useful design features of a fluid reservoir.

Figure 11.3 illustrates a number of constructional details which should preferably be incorporated into the fluid reservoir. The low pressure return line to the reservoir should be positioned well below the fluid level. By this means unnecessary agitation of the fluid is avoided. Furthermore the tank should be designed such that the returned fluid has time to release air bubbles before again entering the suction line to the pump. This can pose problems in modern, compact systems, though sensible use of baffle plates to delay the fluid's arrival at the pump intake will allow more time to separate from contaminants. Incorporation of dividing baffles in the form of weirs can assist in localizing turbulence near the return line and trapping sedimenting mechanical contaminants before reaching the suction line. A wire gauze screen (stainless steel, approximately 20 mesh/cm) placed at an angle of $\leq 30°$ between the return and suction lines, can effectively remove entrained air from the circulating fluid. It is, however, imperative that the screen be totally immersed in the fluid as surface foam is capable of penetrating even the finest of screens.

Fluids of low viscosity release entrained air more rapidly than similar fluids of higher viscosity (Fig. 11.4). Some few parts per million of high molecular weight polysiloxanes ('silicone oil') are normally added to hydraulic oils to suppress the formation of surface foam in the oil reservoir. These additives are only effective against

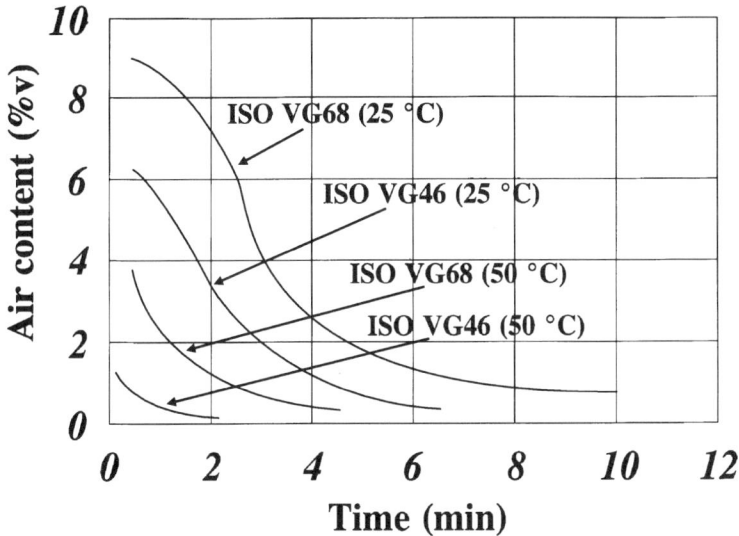

Fig. 11.4 Effect of viscosity and temperature on air release.

surface foam, and there is no purpose in adding further amounts of anti-foamant to counteract air entrainment in a system drawing in false air via faulty seals, pipe connections, etc. An excess of conventional surface-active defoamant additives will thus aggravate the situation by stabilizing the finely dispersed air bubbles and delay their ascent to the oil surface (Fig. 11.5).

Accidental contamination by silicone materials may occasionally originate from certain sealing pastes or mould-release oils utilized during the manufacture of hydraulic hoses, etc.

11.1 Diagnosis and treatment of aeration problems

A correct assessment of the cause(s) is required in order to solve operational irregularities in a quick and economic manner. This is especially important in connection with foaming and air entrainment as incorrect remedies can, as indicated above, intensify the difficulties.

The system should first be checked for possible mechanical defects, incorrect oil level, etc. If no obvious faults are detected, samples of the fluid should be be tested with respect to

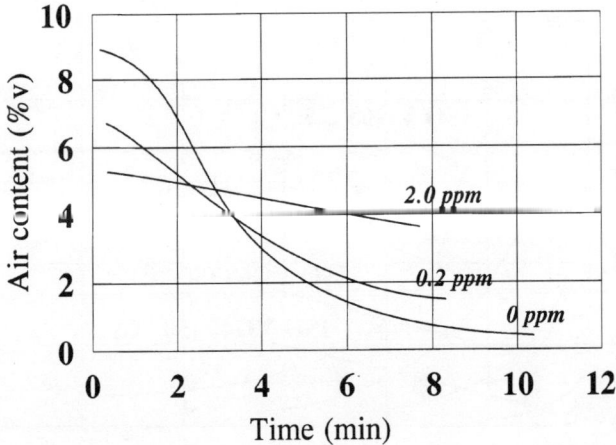

Fig. 11.5 Effect of silicones on air release at 25°C.

Table 11.1 Probable causes of aeration problems

Analytical result		
Air release (IP 313)	Foaming tendency (ASTM D892)	Probable cause
Rapid	Low	Mechanical fault
Slow	Low	Contamination by silicones
Slow	High	Contamination by basic materials, e.g. engine oils, detergents, etc.

(a) air release by test method IP 313,
(b) foaming properties by test method ASTM D892.

A comparison of the test results wil indicate the probable cause of the operational problems (Table 11.1). Unfortunately no additive treatment or filtration can correct unsatisfactory air release due to dissolved impurities. If, on the other hand, difficulties are caused by particulate contaminants, filtration by means of a suitable microfilter (e.g. 5–10um absolute) should alleviate the problem.

12

FILTERABILITY

Modern high pressure hydraulic equipment incorporating complex arrangements of sensitive servo-valves, pumps, etc., makes considerable demands on the cleanliness aspect of hydraulic fluids, and there is a growing awareness among maintenance personnel of the significant cost savings attainable by due attention to careful maintenance routines and the use of reliable filters. In the exacting field of numerically controlled (NC) machine tools, robotics and aerospace applications incorporating ultra-fine system filters, it is absolutely vital to maintain high cleanliness ratings at all times, in order to ensure truble-free operation.

Although a hydraulic fluid may appear clear and bright, without any sign of visible mechanical impurities, it may nevertheless still contain excessive amounts of microscopic particles capable of affecting the accuracy of critical NC machines. The accuracy of these machines is dependent upon an extremely clean hydraulic fluid as the controlling servo-valves are precision units with very fine clearances. Any content of hard particles in the fluid could easily damage the sharp metering edges of the valves by abrasion or erosion, thus resulting in machining inaccuracies. A remarkable degree of erosive wear can occur on even hardened and polished surfaces adjacent to high velocity fluid streams when mechanical impurities are present. Typical examples of this are the damaged servo-valves from fighter aircraft, where valve openings are extremely small and critical for the necessary accuracy in operation.

The increased appreciation of potential maintenance savings and improved system reliability to be gained by attention to cleanliness aspects, has resulted in increasing demands for cleanliness guarantees from suppliers of hydraulic fluids. This has, for example, become a standard specification requirement in most North Sea offshore supply contracts. Nevertheless, it is generally recognized that ultra-fine filtration of bulk supplies at the point of manufacture is hardly practical as subsequent contamination during transport and handling would nullify the effect and the cost would be prohibitive. A far more preferable approach is to ensure that production and supply procedures are such as to ensure an acceptable cleanliness level, monitored by regular particle counts, and pre-filter charges via a suitable micro-filter directly into critical systems as required.

The presence of microscopic mechanical impurities in the fluid is only one aspect of filterability. Another less obvious, though equally troublesome, problem is the tendency of certain additives and additive combinations to block fine filters by deposition of gelatinous material on the filter surface. Attention was first drawn to this problem in the Nordic countries during low temperature operation of forestry machinery, the onset of filter blocking being immediately obvious as machines were started from cold with relatively viscous oil. Not surprisingly, this phenomenon is accentuated by the presence of water, and filterability test procedures usually take this factor into account. The gelatinous deposits appear to be derived from several possible sources, including hydrolysis of zinc additives and metal soaps formed by reaction between acidic rust inhibitors and basic components, e.g. calcium or zinc compounds. Although many hydraulic oil formulations include small amounts of calcium additives, filter blocking problems in the past have frequently been traced to contamination of the fluid by highly basic engine oils. On account of the wide variety of additive systems in use, blending different fluid types may often result in significantly poorer filterability. Other possible disadvantages of blending dissimilar formulations are the risk of increased air entrainment, foaming and generally reduced performance due to additive interaction. It is therefore recommended to avoid mixing different formulations of hydraulic fluids unless the proposed blend has previously been screened by laboratory testing.

12.1 Filterability test procedures

A number of standardized and ad hoc 'in-house' test methods are used to investigate the filterability of hydraulic fluids.

ASTM F52–69 'Silting index of fluids for processing electron and microelectronic devices' was developed for the aerospace industry and specifies filtration of the test fluid through a 0.8 µm membrane filter at a high constant pressure differential. While particles larger than 5 µm tend to form a loose permeable cake, particles smaller than 5 µm and gelatinous materials tend to block the porosity of the filter and cause a decay in the rate of flow. The flow rate is reported as a dimensionless *silting index*.

The UK Ministry of Defence specification for OM-33 (Nato H-576) hydraulic oils specifies a similar test procedure, using a filter of 5 µm porosity instead of 0.8 µm. The silting index of an unused hydraulic oil, examined by this procedure, will normally be in the region of 0.2–0.8 (the OM-33 limit is ≤1.0). Any tendency of the oil to form gelatinous deposits when contaminated by moisture or basic additives results in a dramatically reduced rate of filtration, yielding extremely high silting indices.

The *AHEM* procedure involves measuring the time for 300 ml of test fluid to pass through, or block, a membrane filter of 1.2 µm porosity and 9.6 cm cross-sectional area.

The *CETOP Filtration Test G6.15.01D*, favoured, among others, by major consumers in the Nordic countries, resembles the AHEM procedure, but is more critical, as 1000 ml of fluid is filtered through the 1.2 µm membrane. Another difference is the calculation of a filter factor (FF) obtained by dividing the volume of oil filtered prior to blocking by the area of the filter membrane. The maximum filter factor possible is thus 1000/9.8 = 104. Susceptibility of the test formulation to contamination by water may be evaluated by repeating the test after addition of 1% water to the test fluid.

Denison TP 02100 is the required filterability test procedure in the American Denison HF-0 specification for hydraulic oil with anti-wear additives for severe duty. In recognition of the deleterious effect of water contamination on the filterability of many oil formulations, this manufacturer specifies maximum filtration times for the test fluid through a 1.2 µm membrane, with and without 2% water. A filtration time of 600 s is allowed for the fluid itself, and with 2% water the filtration time must not exceed 1200 s.

13

SPECIFICATIONS

As may be expected, the majority of manufacturers of hydraulic equipment recommend certain performance standards for hydraulic fluids used in their components. In some instances recommendations are confined to viscosity grades, whilst in other cases very comprehensive quality requirements are specified, e.g. oxidation stability, filterability, anti-wear properties, air-release, etc.

Among important specifications the following may be mentioned:

1. The German DIN 51 524 (part 2, H-LP and part 3, HVLP), specifying technical requirements for anti-wear oils. The specification ensures satisfactory anti-wear performance for heavily loaded steel–steel contacts by a minimum failure load stage requirement of 10 in the FZG gear test rig.
2. Vickers M-2952-S, describing hydraulic oils for mobile equipment. The standard specifies viscosity limits and requires approval by the exacting 35VQ25 pump test.
3. Denison (Hägglund) HF-0, notable for requiring high thermal stability by the Cincinnati-Milacron P-75 test (168h at 135°C) together with hydrolytic stability by ASTM D2619. The HF-0 test was developed as a more severe version of the earlier HF-2 test following wear problems with bronze components in axial piston pumps.
4. Swedish Standard SS 155434, requiring good low temperature flow properties and satisfactory performance in hydraulic systems outdoors under Nordic climatic conditions.
5. NATO code H-515 (MIL-H-5606 or DEF STAN 91/48), 'superclean'

mineral hydraulic oil possessing a high viscosity index (>300) and extremely good flow properties at low temperatures.
6. NATO code H-537 (MIL-H-83282), synthetic hydrocarbon-based hydraulic fluid with improved fire-resistant properties compared to mineral oil fluids.

Certain specification requirements are seemingly incompatible, e.g. chemically active anti-wear additives and high thermal stability, and only during recent years have products been formulated capable of satisfying the joint quality requirements of specifications 1–3 mentioned above.

13.1 Requirements

The requirements are listed in Tables 13.1–13.8 presented on the following pages.

Table 13.1 German specification for industrial hydraulic fluids, DIN 51 524 – parts I and II

	DIN	HL						HLP					
ISO grade		10	22	32	46	68	100	10	22	32	46	68	100
Viscosity, cSt at 40°C (Min.)	DIN 51 519	9	19.8	28.8	41.4	61.2	90	9	19.8	28.8	41.4	61.2	90
at 40°C (max.)	DIN 51 550 &	11	24.2	35.2	50.6	74.8	110	11	24.2	35.2	50.6	74.8	110
at 100°C (min.)	DIN 51 562	2.4	4.1	5.0	6.1	7.8	9.9	2.4	4.1	5.0	6.1	7.8	9.9
at 0°C (max.)		90	300	420	780	1400	2560	90	300	420	780	1400	2560
at −20°C (max.)		600						600					
Flash point, °C (min.)	DIN 51 376	125	165	175	185	195	205	125	165	175	185	195	205
Pour point, °C (max.)	DIN 51 597	−30	−21	−18	−15	−12	−12	−30	−21	−18	−15	−12	−12
Air release, min. (max.)	DIN 51 381	5	5	5	10	10	14	5	5	5	10	10	14
Demulsibility, s (max.)	DIN 51 599	30	40	40	40	60	60	30	40	40	40	60	60
Foam test ml, (max.)	DIN 51 566	150/0 – 75/0 –150/0						150/0 – ⁻5/0 – 150/0					
Copper corrosion, 3h/100°C	DIN 51 759	2						2					
Steel corrosion	DIN 51 585	max. 0 – A						max. 0 – A					
Seal compatibility (SRE-NBR1, 7 days/100°C)	DIN 53 521												
(a) Volume change (%)	DIN 53 505	0–18	0–15		0–12	0–10		0–18	0–15		0–12	0–10	
(b) Shore A hardness change		0/–10	0/–8		0/–7	0/–6		0/–10	0/–8		0/–7	0/–6	
FZG test	DIN 51 354												
(a) Damage stage (min.)											10		
(b) Wear mg/kWh, (max.)											0.27		
Vickers vane pump	DIN 51 389												
(a) Wear of ring (mg) (max.)										120			
(b) Wear of vanes mg, (max.)										30			
Oxidation test Neutralization value after 1000h (mg KOH/g) (max.)	DIN 51 558	2.0						2.0					

Table 13.2 German specification DIN 51 524 – part III

				HVLP		
ISO grade	DIN 51 519	15	32	46	68	100
Viscosity, cSt at 40°C (min.)	DIN 51 550	13.5	28.8	41.4	61.2	90
at 40°C (max.)	&	16.5	35.2	50.6	74.8	110
at 100°C (min.)	DIN 51 562	As advised by supplier				
at 0°C (max.)		As advised by supplier				
at −20°C (max.)		As advised by supplier				
Kinematic viscosity index	ISO 2909	≥ 140				≥ 120
Shear stability, 250 cycles – viscosity loss at 40°C (%)	DIN 51 382	As advised by supplier				
Flash point, °C (min.)	ISO 2592	125	175	180	180	190
Pour point, °C (max.)	ISO 3016	−39	−30	−217	−24	−21
Air release, minutes (max.)	DIN 51 381	5		10		14
Demulsibility, s (max.)	DIN 51 599	30	40		60	
Foam test, ml (max.)	DIN 51 566	150/0 − 75/0 − 150/0				
Copper corrosion, 3 h/100°C	DIN 51 759	2				
Steel corrosion	DIN 51 585	max. 0 − A				
Seal compatibility (SRE-BNR1, 7 days/100°C)						
a) Volume change (%)	DIN 53 521	0–15	0–12		0–10	
b) Shore A hardness change	DIN 53 505	0/−8	0/−7		0/−6	
FZG test	DIN 51 354					
a) Damage stage (min.)		10				
b) Wear (mg/kWh) (max.)		0.27				
Vickers vane pump	DIN 51 389					
a) Wear of ring (mg) (max.)		120				
b) Wear of vanes (mg) (max.)		30				
Oxidation test Neutralization value after 1000 hours (mg KOH/g) (max.)	DIN 51 558	2.0				
Particle content (%m)	DIN 51 592	Not measurable				
Water content (%m)	ISO 3733	Not measureable				

Table 13.3 Vickers specification requirements for mobile and industrial equipment

	M–2950–S	I–286–S
Pump tests		
35VQ25 vane pump	Yes	—
V–104C vane pump	—	Yes
Oxidation tests		
ASTM D943 (hours to TAN=2,0 mg KOH/g)	*	*
ASTM D4610 (1000 hours test)	*	*
CIGRE	*	*
ASTM D2272 (RBOT)	*	*
Rust prevention		
ASTM D665A	*	*
ASTM D665B	*	*
Thermal stability		
168 hours at 135°C	*	—
Viscosity		
ISO VG	**	**

*   Satisfactory performance must be documented, limits not specified.
**   Suitable viscosity grade for the application.

Table 13.4 Denison specifications HF–O and HF–2 for anti-wear hydraulic oils

		HF–2	HF–0
Viscosity index (min.)	ASTM D2270	90	90
Aniline point (°C) (min.)	ASTM D611	—	100
Rust prevention	ASTM D665	Pass A & B	Pass A & B
Foam test Allowable foam after 10 min	ASTM D892	No foam	No foam
Filterability (1.2 µm) a) without water b) with 2% water	Denison–TP 02100	Not required — —	 Max. 600 s Not to exceed 2 × time without water.
Thermal stability a) Sludge (mg) (max.) b) Copper weight loss (mg) (max.) c) Copper rod rating	Cincinnati–Milacron P–75	Not required — — —	 100 10 Report
Oxidation test (1000 h) a) Total acidity (mg KOH/g) (max.) b) Sludge insolubles (mg) (max.) c) Copper weight loss (mg) (max.) d) Iron weight loss (mg) (max.)	ASTM D943	 2.0 400 200 100	 2.0 200 50 50
Vane pump test (100 h) Total weight loss (mg) (max.)	ASTM D288	 50	Pass —
Denison pump tests a) High pressure vane pump (21 MPa) b) Axial piston pump (34.5 MPa)	 T5D 042 P 46	Not required — —	 Pass Pass

Table 13.5 Swedish standard SS 15 54 34 (hydraulic oils)

Property	Requirement SV 15	SV 68	SIS 32	SIS 46	Method
Kinematic viscosity at 100°C (mm²/s)		Report			ISO 3104
Viscosity at 100°C after shearing, 250 cycles (mm²/s)	>3.2	>8.5	>6.0	>9.0	CEC L–14–A–78
Viscosity loss (max.) (%)	10	15	10	10	
Kinematic viscosity At −20°C (mm²/s) At −30°C (mm²/s)	— ≤1800	≤5000 —	— ≤4000	≤2400 —	ISO 3104
Pour point (°C)	≤ −39	≤ −30	≤ −39	≤ −39	ISO 3106
Flash point (min.) (°C)	140	180	140	140	SIS 02 18 12
Air release value at 50°C (max.) (min)	5	12	10	10	DIN 51 381
Vickers vane pump test Weight loss (cam ring) (mg) Weight loss (vanes) (mg)		⟵— Max. 120 ———⟶ Max. 30			IP 281
Rust prevention (24 h at 60°C)		⟵— Pass ————⟶			ASTM D665A
Water separability, (oil/water/emulsion) (ml)		⟵— 40/40/0 ———⟶			ASTM D1401
Water & solid particles		⟵— No traces ——⟶			Visual

Table 13.6 Specification requirements, MIL–H–5606F (extract)

Property	Requirement	Test method
Base oil	Mineral	—
Flash point (PMCC) °C	min. 82	ASTM D93
Kinematic viscosity, mm²/s		ASTM D445
At −40°C	max. 600	
−54°C	max. 2500	
40°C	min. 13.2	
100°C	min. 4.90	
Pour point, °C	max. −60	ASTM D97
Neutralization no., mg KOH/g	max. 0.20	ASTM D664
Water content, ppm	max. 100	ASTM D1744
Colour	Red	ASTM D1500
Mechanical impurities, particles/100 ml		Federal STD–791
of size (μm): 5–15	max. 10000	method 3009
16–25	max. 1000	
26–50	max. 150	
51–100	max. 20	
>100	max. 5	
Copper corrosion test	max. 2e	ASTM D130
Lubrication, 4-ball test:		ASTM D4172
wear scar (40 kg load) mm	max. 1.0	
Elastomer compatibility,		
NBR-L rubber swell, %	8–30	
Corrosion-oxidation stability,		Federal STD–791
168 h at 135°C:		method 5308
Viscosity change at 40°C, %	−5/+20	
Acidity increase, mg KOH/g	0.20	
Metal weight change (mg/cm²)		
Aluminium	≤ 0.2	
Magnesium	≤ 0.2	
Cadmium	≤ 0.2	
Steel	≤ 0.2	
Copper	≤ 0.2	

Table 13.7 Specification requirements, MIL–H–83282C (extract)

Property	Requirement	Test method
Base oil	Synthetic hydrocarbon	—
Flash point (COC) (°C)	min. 205	ASTM D93
Fire point (COC (°C)	min. 245	ASTM D92
Autoignition temperature (°C)	min. 345	ASTM E659
Kinematic viscosity, mm^2/s		ASTM D445
At −40°C	max. 2200	
−54°C	—	
40°C	min. 14.0	
100°C	min. 3.45	
Pour point (°C)	max. −55	ASTM D97
Neutralization no. (mg KOH/g)	—	ASTM D664
Water content (ppm)	max. 100	ASTM D1744
Colour	Red	ASTM D1500
Mechanical impurities, particles/100 ml		Federal STD–791 method 3009
of size (μm): 5–15	max. 10000	
16–25	max. 1000	
26–50	max. 150	
51–100	max. 20	
>100	max. 5	
Copper corosion test	max. 2e	ASTM D130
Lubrication, 4-ball test:		ASTM D4172
wear scar (40 kg load) (mm)	max. 0.65	
Elastomer compatibility, NBR-L rubber swell (%)	8–30	
Corrosion-oxidation stability, 168 h at 135°C		Federal STD–791 method 5308
Viscosity change at 40°C (%)	≤ 10	
Acidity increase (mg KOH/g)	≤ 0.20	
Metal weight change (mg/cm^2):		
Aluminium	≤ 0.2	
Magnesium	≤ 0.2	
Cadmium	≤ 0.2	
Steel	≤ 0.2	
Copper	≤ 0.2	

Table 13.8 MIL–H–8446B hydraulic fluid, non-petroleum base, aircraft

Property	Requirement	Method
Viscosity, cSt at		ASTM D445
−54°C (max.)	2500	
204°C (min.)	2.5	
Pour point °C, (max.)	−59	ASTM D97
Vapour pressure at 204°C mm Hg, (max.)	5	ASTM D2879
Autogenous ignition temperature °C, (min.)	371	ASTM D286
Neutralization no. (max.)	0.2	ASTM D664
Rubber swell, standard 'R' stock	25 ± 5%	Fed. Std. 791/3603
Foaming at 93° C		Fed. Std. 791/3212
Total volume oil + foam ml, (max.)	600	
Collapse time s, (max.)	600	
Low temperature stability at −54°C	No gel or solidification	
Oxidation-corrosion test at 204°C:		Fed. Std. 791/5308
Weight loss of metals (mg/cm^2)		
Copper	±0.4	
Aluminium, steel, silver	±0.2	
Viscosity change at 99°C	±20%	
Appearance of metals after test	No pitting or corrosion	
Neutralization no. change (max.)	1.0	
Oil appearance	Clear	
Hydrolytic stability at 93°C:		Fed. Std. 791/3457
Weight change – copper	0.5	
Appearance of copper	No pitting	
Acid no.		
oil phase change (max.)	0.5	
aqueous phase (max.)	0.5	
Viscosity change at 99°C (max.)	±20%	
Insolubles (max.)	0.5	

14

HYDRAULIC FLUIDS FOR MILITARY AND AEROSPACE APPLICATIONS

Military and aerospace applications utilize a broad spectrum of hydraulic fluids, ranging from simple straight mineral oils to highly specialized synthetic fluids. Understandably enough, this area of application is dominated by a high degree of interest in fire-resistant media with widespan temperature capabilities. Products of this nature are increasingly specified for use in military aircraft, naval vessels, combat vehicles and weapon systems.

A selection of the most well-known military specifications are listed in Table 14.1.

14.1 Aircraft and aerospace

Significant advances in system and component design within the aerospace industry have been accompanied by correspondingly exacting demands on hydraulic fluids required to operate in these systems. Manual controls have been replaced by hydraulics and the number of hydraulically operated functions has increased rapidly, as has the complexity of the circuits. Operating parameters have also become more critical; high altitudes require improved low temperature fluidity whilst higher pressures (≈ 20 MPa) and reduced fluid volumes necessitate better lubrication properties, and thermal and oxidation stability.

Although the use of phosphate ester fluids in commercial aircraft has been well established since their introduction during the 1950s, it is only in the course of the past decade that the widespread use of mineral hydraulic oils (Nato code H-515) in military aircraft has noticeably decreased. On account of the comprehensive system modifications required, and the operational temperature limitation (≈110°C), NATO aircraft were not converted to use phosphate ester products.

However, the search for suitable fire-resistant fluids has continued, and a silicate ester fluid was finally defined by the new specification MIL-H-8446, characterized by extreme temperature capabilities (−54°C to +204°C), high thermal stability and good fire resistance. This fluid was based on alkyl silicate esters with the addition of ethylmethyl siloxane to improve the viscosity index and regulate seal swell. Unfortunately, relatively poor hydrolytic stability detracted somewhat from the otherwise excellent performance properties, and MIL-H-8446 is now obsolete. Nevertheless, a similar type of silicate ester fluid is still utilized in the central hydraulic system of the supersonic Concorde aircraft.

The high incidence of hydraulic fluid fires in US helicopters hit by enemy gunfire in the Korean war highlighted the need for a more fire-resistant fluid. In 1974 a replacement for the mineral oil-based MIL-H-5606 (H-515) products was developed in the form of a synthetic hydrocarbon fluid, later designated MIL-H-83282 (H-537). One of the primary advantages of this fluid over phosphate esters is complete compatibility and interchangeability with conventional mineral oil-based MIL-H-5606 fluids, thus permitting use in aircraft systems designed for these media. The new fluid is slightly inferior to the original mineral oil with respect to low temperature fluidity, but offers considerably improved high temperature properties and lower flammability (see Table 14.2).

No single flammability test correlates fully with the many and varied conditions prevailing in real situations, and it is therefore usual to examine candidate fluids by several test methods to form a reliable assessment of their fire-resistant potential. In Table 14.2, it will be noted that although the synthetic hydrocarbon fluid displays significantly higher test values with respect to flash point, fire point and autoignition temperature, this is not the case for the hot plate ignition test. This apparently anomolous result is due to the higher volatility of the MIL-H-5606 fluid and shorter residence time at the hot surface.

The new synthetic hydrocarbon fluid was initially utilized by US Navy and Army helicopters. Subsequent experience revealed that the low temperature limitation of the new MIL-H-83282 fluids was essentially a ground-level cold-soak problem, and many types of military aircraft have since been successfully converted to the synthetic hydrocarbon fluid.

Despite the improved performance characteristics of MIL-H-83282

Table 14.1 Hydraulic fluids for military applications

Nato code	Military specification	Application	Fluid type	Viscosity (mm²/s)			Pour point °C	Flash point °C
				100°C (min.)	40°C	−40°C (max.)		
H-515	MIL-H-H606	Hydraulic systems on civil and military aircraft	Mineral oil	4.90	≥13.2	600	−60	82
H-536	MIL-S-83282	Lubrication of gyrocompass and rocket systems over a wide temperature range	Chlorophenyl silicone oil	15/19	50/60	—	−75	290
H-537	MIL-H-83282	Aircraft, weapons and other hydraulic systems with synthetic seals	Synthetic hydrocarbon	3.5	≥14	220	−55	205
H-542	SAE J 1703	Brake fluid for military administrative vehicles and certain submarine systems	Glycol ether and polyglycol	1.5	3.5 (50°C)	900 (−35°C)	—	82
H-544	MIL-H-46170	Combat vehicles recoil oil for artillery, aiming mechanisms for guns and tanks	Synthetic hydrocarbon	3.4	≦19.5	260	−54	218
H-547	MIL-B-46176	Brake fluid for combat vehicles	Silicone oil	1.3	—	900 (−55°C)	—	—

H-572	BS 4475/1975 grade CSB-68 (OM-65)	Gun mechanisms, aiming devices, gyrocompass and general lubrication at moderate temperatures	Mineral oil	—	61/75	—	−9	192
H-573	MIL-L-17672	Mooring winches, steering gear, submarine central hydraulic systems	Mineral oil	—	41.4/50.6	—	−23	163
H-575	MIL-F-17111	Power transmission fluid, naval ordnance equipment	Mineral oil	8.0	25	—	−40	104
H-576	DEF STAN 91/39 (OM-33)	Hydraulically operated rockets, steering gear, etc.	Mineral oil	—	26/33	—	−30	160
H-579	MIL-H-22072	Catapult systems onboard aircraft carriers	Water/glycol	—	38/43	—	—	—
H-580	MIL-H-19457	Catapult systems and deck elevators on aircraft carriers	Phosphate ester	4.8	38.5/45/5	—	−18	—

Table 14.2 Relative flammability characteristics of petroleum- and synthetic hydrocarbon-based aviation hydraulic fluid (typical data)

Property	MIL-H-5606	MIL-H-83282
Base oil type	Mineral	Polyalphaolefin
Flash point (COC) (°C)	103	220
Fire point (COC) (°C)	115	254
Hot plate ignition test (°C)	590	540
Autoignition temperature (°C)	232	371
Incendiary gunfire test, (50 calibre/ 4 shots) number of fires	5	1

fluids, the principal deficiency associated with this fluid and phosphate esters is that under many operational conditions, both media may still ignite and burn, thus only providing a degree of improvement with respect to the frequency and intensity of hydraulic system fires.

Silicone oils, modified by use of appropriate additives, have also shown promise as fire-resistant aircraft hydraulic fluids. Other inherent properties such as excellent low temperature flow properties, extremely high viscosity indices, low volatility, good hydrolytic stability and high oxidation stability would seem ideal for this application. However, these advantages are somewhat overshadowed by poor boundary lubrication properties for steel-on-steel and relatively high compressibility compared to mineral oil and synthetic hydrocarbon-based fluids. Nevertheless, a fluid based on chlorophenylmethylsiloxane, was tested in BAC TSR2 and Hawker Hunter FGA Mk6 aircraft, modified with respect to seal materials. The trials were reasonably satisfactory, but as further modifications involving servo-valve design were considered desirable, adoption of silicone fluids was not pursued.

Silicones, phosphate esters and a variety of halogenated compounds have all been rejected by military research laboratories in the search for the ultimate fire-resistant hydraulic fluid. Among the most promising base materials mentioned to date are chlorotrifluoroethylene, fluoroalkylether, perfluoroalkylether and perfluoroalkylether triazine.

14.2 Combat vehicles and artillery

Earlier hydraulic fluids utilized by NATO forces were almost exclusively mineral oil products, e.g. H-540 or C-635 for vehicles and artillery. During the Israeli–Arab conflict in 1973, considerable losses were experienced due to fires when the hydraulic systems of combat vehicles were hit by enemy gunfire. Consequently, the US Army intensified research on fire-resistant media and developed a corrosion-inhibited version of the synthetic hydrocarbon MIL-H-83282 fluid. The new fluid is standardized in accordance with MIL-H-46170 (Nato-code H-544), and is now widely applied within NATO, principally for the hydraulic systems of combat vehicles.

For many years military vehicles utilized brake fluids based on polyglycols and glycolethers, different products being specified for various climatic conditions and storage purposes. The water sensitivity of these products is recognized as a major factor in brake system failures following ingress of moisture into the system via rubber seals and hoses. The absorbed water lowers the boiling point and vapour lock point of these fluids, and significantly increases low temperature viscosity, thus adversely affecting their performance. In order to avert these problems, non-hydroscopic silicone fluids have been developed. These fluids possess high boiling point, high viscosity indices and favourable low temperature viscosity. On account of the inherent low surface tension of these fluids, they display excellent wetting ability for metal surfaces, thereby hindering penetration of moisture and subsequent corrosion. On account of the insensitivity of silicone brake fluids to moisture, and their good chemical stability, the previous requirement for periodical replacement of the fluid has been largely eliminated.

14.3 Naval vessels

Hydraulic fluids used in naval vessels are at present mainly based on conventional mineral oil fluids, e.g. Nato codes H-572, H-573 and H-576. There is naturally also here a distinct requirement for fluids offering improved fire resistance. During the 1960s a fluid based on tertiary-butyl triphenyl phosphate had already been developed by the US Navy for use in catapult launching systems and flight-deck elevators on board aircraft carriers. This fluid is standardized in accordance with MIL-H-19457 (H-580).

Following experience during the Falklands war, the Royal Navy increased efforts to develop more suitable hydraulic fluids, culminating in a new water–glycol fluid with, among other things, improved low temperature properties compared to H-579 (MIL-H-22072).

A significant problem associated with hydraulic system fires is the formation of toxic gases, which represents a very serious danger where some of the otherwise effective fire-resistant products are concerned, e.g. phosphate esters and halogenated compounds. This problem is particularly acute on board submarines. Another important requirement with respect to submarine operations is the preferred use of hydraulic fluids miscible with sea water, thus ensuring leakages do not reveal the position of the vessel by local fluid concentrations on the surface.

15

SELECTION OF A SUITABLE HYDRAULIC FLUID

In order to ensure a reliable choice of suitable hydraulic fluid, it is essential to start by compiling information as detailed as possible concerning the actual system and operating conditions. The most important facts to consider are as follows.

- What are the upper and lower viscosity limitations?
  (See also Chapter 6.)
- Over what temperature range will the system operate?
- Should a fire-resistant fluid be used?
- Are there special lubrication requirements concerning any of the components?
- Is the choice of fluid types restricted by any of the construction materials, e.g. seals or filter elements?
- What are the manufacturer's recommendations with respect to the most critical components (especially hydraulic pumps and motors).
- Are there other special requirements, e.g. environmental constraints?

After careful consideration of the above factors, the optimal viscosity grade of media fulfilling the remaining technical criteria is determined. If necessary, the final selection is made after also considering possible modification of the system to accommodate a preferred fluid type, for technical, economic or rationalization reasons.

In general, the majority of industrial applications are satisfied by an ISO-type HM mineral hydraulic oil of viscosity grade ISO VG32 or 46. Systems operating at relatively high temperatures, for example plastic moulding machines, often utilize the more viscous grades, ISO VG68 or even 100.

Mobile equipment normally utilizes ISO-type HV, ISO VG32 mineral hydraulic oil over a temperature range of $-20°C$ to $+70°C$. Ambient temperatures below $-20°C$ require the use of mineral or synthetic fluids possessing higher viscosity indices and good shear stability. Especially severe requirements are found in connection with certain types of forestry equipment operating under low ambient temperatures and arduous conditions, as in the Nordic countries for instance. Many of these machines incorporate hydraulic motors running under slow speed/high torque conditions, thus necessitating the use of highly shear-stable ISO-type HV, ISO VG46–68 products.

For systems in which the normal temperature of operation exceeds $90°C$, use of a suitable synthetic fluid possessing adequate oxidation stability should be considered. As mentioned in previous chapters, particular attention must be directed towards compatibility with surface coatings, seals and other eventual elastomeric materials when selecting a synthetic fluid.

16

TEST METHODS FOR HYDRAULIC MEDIA

The majority of standardized test methods are issued by the following:

ASTM (American Society for Testing and Materials)
IP (Institute of Petroleum)
DIN (Deutsches Institut für Normung)
FTM (Federal Test Method)
SIS (Standardiseringskommissionen i Sverige)
ISO (International Organization for Standardization)
AHEM (Association of Hydraulic Equipment Manufacturers)

In addition there are many 'in-house' procedures developed by pump manufacturers, oil companies, etc.

16.1 Physical–chemical properties

Kinematic viscosity: ASTM D445, IP 71   Viscosity is a measure of a fluid's internal friction, or resistance to flow when subjected to a shearing force. The method determines the kinematic viscosity of the fluid by measuring the time for a predetermined volume of the liquid to flow under gravity through a calibrated glass capillary viscometer. The unit of measurement is the centistoke (cSt) or mm/s, and may be converted to dynamic viscosity in centipoise (cP = mPa·s) by multiplying the kinematic viscosity and density of the fluid. As in all viscosity measurements, exact temperature control is of the utmost importance to avoid inaccuracies.

Pour point: ASTM D97, IP 15 The pour point is the lowest temperature, expressed as a multiple of 3°C, at which the test fluid is observed to flow when cooled and examined under prescribed conditions. As a general rule the pour point should be at least 10°C below the lowest temperature at which the fluid is to be used. It is, however, important to appreciate that it is the actual viscosity at this temperature, and not the pour point itself, that is the real limiting factor for low temperature operation.

Flash point: ASTM D92 or D93 The flash point is the temperature to which a fluid must be heated in standard test apparatus to evolve sufficient vapour to flash momentarily on application of a small flame. The test may be performed either as an 'open cup' test (D92), or a 'closed' test (D93). The latter procedure is more sensitive to small amounts of volatile constituents and will always yield lower values (15–30°C) than the open cup version.

ASTM D92 describes the determination of the flash point according to the Cleveland Open Cup method.

ASTM D93 describes the Pensky-Martens Closed Cup test procedure. This is the preferred method when contamination of the hydraulic fluid by volatile fractions is suspected, but it is also often used for convenience on account of the many automatic testers available.

Neutralization number, acid value or TAN: ASTM D664, IP177 This method describes procedures for the determination of acidic or basic constituents in petroleum products. The method resolves these constituents into groups having weak-acid, strong-acid, weak-base or strong-base ionization properties.

The acidity of unused oils and fluids is normally derived from the type and concentration of specific additive materials. This is particularly relevant to fluids containing metal dithiophosphate additives, these products being characterized by a relatively high initial acidity (≤1.5 mg KOH/g).

The acidity is otherwise of interest as a measure of the degree of oxidation of a used fluid, and by monitoring the development of acidity over time, it provides an indication of the useful life of the fluid.

In ASTM D664 and IP 177 the oil is dissolved in an ionizing solvent and titrated potentiometrically with potassium hydroxide. A submerged electrode registers changes in the hydrogen ion concentration during the titration, thus detecting the exact neutralization point. Acidity may also be determined by colorimetric titration, using a colour indicator (ASTM D974 and IP 1).

Foaming properties: ASTM D892, IP 146 Foaming is usually caused by air gaining ingress to the system at faulty seals or leaks in the suction line. Unsuitable positioning of fluid return lines in the reservoir, the presence of fine particulate contaminants, and low fluid levels will also favour the accumulation of foam. The presence of small air bubbles significantly reduces the bulk modulus of the fluid and can also promote oxidation.

The test is performed by registering the foaming tendency as air is dispersed in the fluid under standardized test conditions, and thereafter the stability of the foam. The test is carried out in three stages at temperatures of 25°C, 100°C, and then again at 25°C.

Air release: IP 313, DIN 51 381 Whilst ASTM D892 examines the tendency of a fluid to accumulate surface foam, IP 313 determines its ability to separate out entrained air bubbles. A favourable result when examined in accordance with IP 313 or DIN 51 381 indicates that the fluid is unlikely to sustain circulation of entrained air in the system.

The test procedure involves recording the time in minutes for finely dispersed air in the oil to decrease to 0.2%v under standardized test conditions, normally at a temperature of 50°C.

Oxidation stability: ASTM D943, ASTM D2272, IP 328 With the exception of water, all hydraulic fluids are subject to chemical deterioration under the influence of heat and oxygen (air), forming various oxidation products. The oxidation reactions are accelerated by the presence of catalyst materials, e.g. dissolved copper compounds, but may be retarded by addition of inhibitor additives (antioxidants).

A wide variety of standardized and 'in-house' test procedures are employed to evaluate the stability and expected service life of hydraulic fluids under oxidative conditions. The majority of these tests are based on exposing the test fluid to oxygen or air at relatively high temperatures and in the presence of catalyst metals, to increase oxidation rates and reduce the testing period. Oxidation stability is assessed by quantitative determination of the oxidation products, oxygen (air) absorption, viscosity changes, etc., after specified intervals of time. Several of the most popular test methods were originally developed for use with turbine oils, but have since proved suitable also for many types of hydraulic fluid.

Examples of useful standardized test methods are:

- TOST (Turbine Oxidation Stability Test), ASTM D943.
- Rotary Bomb Test, ASTM D2272/IP 229.
- Oxidation Stability of Turbine Oils during use, IP 328.

ASTM D943 reacts the oil sample with oxygen in the presence of water and a copper–iron catalyst at 95°C. The test continues until the measured total acid number (TAN) of the oil reaches 2.0 mg KOH/g or above. The number of test hours for the acidity to reach 2.0 mg KOH/g is termed the 'oxidation life' of the oil. Due to the presence of water during the test, misleading results may be obtained when testing fluids of poor hydrolytic stability.

ASTM D2272 oxidizes the oil at 150°C in the presence of water, metallic copper catalyst and oxygen at 620 kPa pressure. The time elapsing before a pressure drop of 175 kPa is registered, is an indication of the oxidation stability. The test procedure is considered useful for production control of oils having the same composition and for monitoring the remaining oxidation life of oils in service. The above remarks concerning fluids deficient in hydrolytic stability also apply to this test procedure.

IP 328 subjects the oil sample to oxidation in an atmosphere of oxygen at a temperature of 150°C, but without catalysts. Oxygen consumed in the oxidation reactions is replaced by an equivalent amount supplied from an electrolytic cell.The time taken for the oil to absorb a specified volume of oxygen is recorded and the result is expressed as the time in hours for 100 g of oil to absorb 300 ml of oxygen.

Thermal stability: ASTM D2160, ASTM D130, C-M test   Thermal stability is really the ability of the fluid to resist chemical change when subjected to high temperature in the absence of air or oxygen (a condition rarely met in practice, air usually being present). Refined mineral base oils possess natural thermal stability up to temperatures around 320°C, and the thermal stability of mineral hydraulic oils is primarily an expression of the additives' resistance to high temperature. The more active dithiophosphate additives are especially prone to thermal degradation, and decompose forming insoluble compounds ('sludge') at high temperatures.

ASTM D2160 evaluates the thermal stability of the fluid at 260°C and 316°C for periods of 6 h in the absence of oxygen and water. Due to potential decomposition and generation of high pressure gas at the test temperature, this method is not suitable for aqueous base fluids or other media generating high pressures.

At the conclusion of the test periods the test cells are rated by visual inspection for gummy deposits, etc., and the sample analysed for viscosity change and neutralization value. The test differentiates the relative stability of hydraulic fluids at elevated temperatures in the absence of air and water under the conditions of the test. In use, thermally unstable fluids may form acidic contaminants causing corrosion, deposits and viscosity changes, leading to system malfunctions.

ASTM D130 describes a simple procedure for investigating the agressiveness of the fluid towards a polished copper strip, at a temperature and for a time selected for the fluid under test. Typical test parameters for non-aqueous fluids are 1–3 hours at 90–150°C. At the end of the test period the copper strip is removed, washed and compared with the ASTM Copper Strip Corrosion Standards. The test clearly differentiates between fluids containing dithiophosphate additives of dissimilar structure.

The most familiar, and perhaps the most important test method for determining thermal stability, is the Cincinnati–Milacron Test (procedure A). In this method the sample fluid is heated to 135°C for a period of 168 h, together with catalyst rods of copper and steel. At the conclusion of the test, the precipitated deposits are weighed, and the weight losses of the catalyst rods are also recorded. This test procedure differentiates very critically between fluids liable to promote wear or corrosion in components incorporating rubbing surfaces of copper alloys against steel, e.g. phosphor-bronze slipper pads in axial piston pumps. The more active forms of anti-wear additives, e.g. *secondary* dialkylphosphates, yield extremely poor results in ths test. The Cincinnati–Milacron thermal stability test is specified in several standards, e.g., in Denison HF-0.

Hydrolytic stability: ASTM D2619   Certain types of base fluid and additives have a tendency to react with water. This results in a deterioration of properties, and some of the products formed during hydrolysis may attack vulnerable metals.

For this test a 75 g sample of fluid plus 25 g of water and a copper test specimen are sealed in a pressure-type beverage bottle. The bottle is rotated end to end, for 48 h in an oven at 93°C. The phases are then separated and insolubles are weighed. The weight change of the copper specimen is measured; the acidity and viscosity changes of the fluid and acidity of the aqueous phase are determined.

The method differentiates the relative stability of hydraulic fluids in the presence of water under the conditions of the test. Hydrolytically unstable hydraulic fluids form acidic and insoluble contaminants which can cause system malfunctions due to corrosion, valve sticking or changes in viscosity of the fluid. This hydrolytic stability test is also specified in the Denison HF-0 standard.

Rust preventive properties: ASTM D665b, ASTM D 1748 Even small amounts of corrosion products can cause catastrophic results in a critical high pressure hydraulic system. Corrosion inhibiting additives are normally incorporated in hydraulic fluid formulations to counteract this eventuality. Various test methods are employed to evaluate the corrosion prevention properties of a fluid, including several oxidation/corrosion tests in which corrosion and formation of surface deposits on metal test strips are assessed under different conditions.

The ability of a hydraulic fluid to prevent rusting in the presence of salt water is examined by means of ASTM D665b. Steel test specimens are submerged in a mixture of the fluid and synthetic sea water at a temperature of 60°C for 24 h. No visible rust is observed with fluids passing the test.

ASTM D1748 may be used to evaluate the rust preventive properties of a non-aqueous hydraulic fluid under conditions of high humidity, correlating to some extent with the conditions in systems contaminated by water, when not in operation. The test is conducted by suspending sandblasted and polished steel specimens, dipped in the test fluid, in a humidity cabinet at a temperature of 50°C and relative humidity of 95–100%. The duration of the test may be varied depending on requirements, the test specimens being rated according to the number and size of rust spots developing during the test period.

Elastomer compatibility: IP 278, ASTM D3604 An essential requirement for hydraulic fluids is that they should be compatible with seal materials and hydraulic flexible hoses. The preferred effect of fluids on hydraulic seals is normally considered to be a moderate swelling (≤2%). On account of the considerable variation in compatibility with different fluid types, it is essential to consider this factor when selecting new materials or fluids.

IP 278 determines the compatibility of mineral hydraulic fluids with nitrile rubber seal materials in terms of a Seal Compatibility Index (SCI). The internal diameter of a standard O-ring is measured on a tapered rod gauge. The ring is then immersed in the oil sample for a period of 24 h at a temperature of 100°C. After cooling the ring for a specified period the change in internal diameter is again measured on the gauge. The Seal Compatibility Index is the percentage volumetric swelling, to the nearest whole number, obtained by conversion from the diametral swell under the specified test conditions. The method can be adapted for use with other elastomers and fluids, and also supplemented by examination of mechanical properties, hardening, etc., following the immersion test.

ASTM D3604 determines the volume change of any selected elastomer in hydraulic fluids, also under static conditions. In this procedure, small test strips of the desired elastomer are immersed in the fluid sample for 168 h at a temperature of 70°C. The volume change of the samples is derived by weighing the sample strips in air and water, prior to and after immersion.

The Swedish standardisation committee have issued their own standard, SIS 162208: *Bestämning av Vätskors Inverkan på Gummi*, involving:

● volume change,
● weight loss,
● change of mechanical properties during immersion,
● change of mechanical properties after solvent extraction.

16.1.12 Water separation: ASTM D1401   Most mineral hydraulic oils and synthetic fluids are designed to separate readily from water, otherwise water contamination could form stable emulsions, increasing viscosity and causing a general deterioration of properties. The resulting emulsion would display increased tendency to disperse solid contaminants, and promote wear, oxidation and corrosion within the system.

ASTM D1401 is a commonly applied test method in which a 40 ml oil sample and 40 ml of distilled water are stirred together for 5 min in a graduated cylinder at 54°C or 82°C. The time for separation of the emulsion thus formed is recorded. When testing synthetic fluids whose relative densities are greater than that of water, the procedure is unchanged, though the separating water phase will probably float on the emulsion or fluid.

16.1.13 Water content: IP 74, ASTM D95, ASTM D1744   IP 74 and the equivalent ASTM D95 are the standard methods for determining water contents of 0.05%v and above in mineral hydraulic fluids and some synthetic fluids. The test procedure involves azeotropic reflux distillation of the oil with xylene, with water separating from the reflux distillate.

ASTM D1744 ('Karl Fischer') is a more accurate method, suitable for water contents between 50 and 1000 ppm. The sample is titrated with a standard solution of iodine, pyridine and sulphur dioxide in methyl alcohol. The end point of the titration, at which free iodine is liberated, may be registered either potentiometrically or by colour indicator.

16.2 Mechanical testing

Anti-wear properties: ASTM D2882, IP 281, Vickers 35VQ25A, Denison P46 Relevant test methods are the pump tests mentioned above, together with a variety of mechanical test rigs designed to simulate various conditions of boundary contact in critical components under a range of different operational conditions.

Table 16.1 lists a number of well-known test rigs and their essential design features.

The major problem with such bench tests is the lack of uniform correlation with the lubricating properties of hydraulic fluids in actual use. Mechanical rig tests tend to be susceptible to specific additive types and good results may, for example, be obtained on FZG or IAE gear rigs by selecting active dithiophosphate compounds. Consequently, the order of merit when rating fluids of dissimilar additive composition will often vary significantly between the different test rigs. The essential criterion when selecting this type of test is similarity of contact conditions between the test rig and the system components (principally pumps) for which the fluid is intended. On account of these considerations, most of the test rigs are utilized primarily by experienced technicians during research and development projects. A notable exception is the FZG test specified in the well-known German national specification DIN 51 524 (parts II and III) for H-LP and HVLP anti-wear hydraulic oils (Fig. 16.1).

Table 16.1 Mechanical test rigs for evaluating anti-wear properties of fluids

Test	Method	Conditions of contact
FZG	IP 334, DIN 51 354	Spur gears, wheel and pinion of steel; loaded line contact
Shell 4–ball	IP 239, ASTM D2266	Steel bearing ball rotating against three fixed bearing balls
Timken	IP 240	Steel ring rotating against steel block
Falex	IP 241	Steel journal rotating against two V-shaped steel blocks
Amsler	'In-house' method	Steel journal rotating against bronze or brass test block

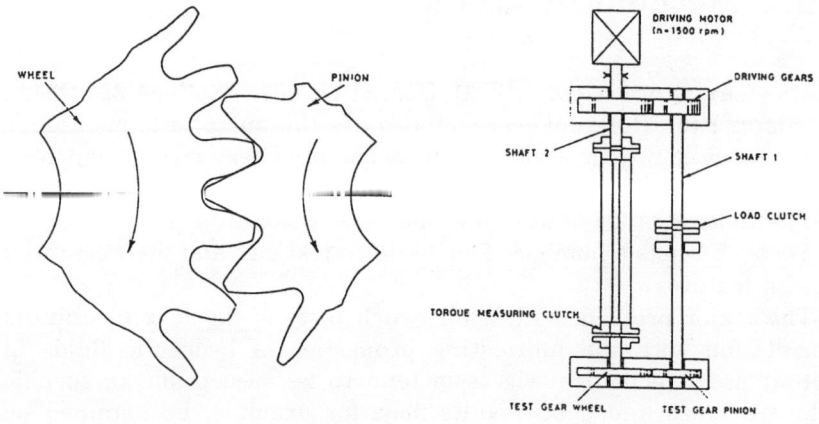

FZG	Test parameters		Test criteria
Type A	8.3 m/s	90°C	Scuffing

Fig. 16.1 Load-carrying capacity by FZG spur gear test rig (DIN 51 354, IP 334).

In the FZG gear rig, used for determining the relative load-carrying capacity of hydraulic fluids and lubricants, the most common wear forms are *scoring* and *scuffing*. The incidence of both these wear patterns is dependent on the composition of the test fluid, and they are consequently used as criteria of performance.

The test is carried out with special spur gears immersed in the test fluid, running at constant speed for 15 min or 21 700 motor revolutions per load stage. At the start of each load stage the fluid temperature is controlled, but allowed to rise freely during the individual stages. The gear loading is increased for each of the specified load stages, and the gears are inspected visually for gear-tooth face damage at the conclusion of each stage. The test is continued until the 'damage load stage' is reached, characterized by the onset of deep scoring lines or seizure areas on any of the gear teeth. The 'type A' test defines 12 load stages, the FZG-value of the test fluid being the above-mentioned damage load stage. If no damage is registered by the final load stage, the test is terminated and the result reported as FZG >12. The 'type B' test, not relevant for hydraulic fluids, is run at a peripheral speed of 16.6 m/s and utilized exclusively for extreme pressure gear lubricants of high load-carrying capacity.

However, it is the behaviour of the hydraulic fluid in practical systems with actual components that is of primary interest. Wear testing performed in full-scale systems would be too extensive, time-consuming and expensive, with many possible inaccuracies. It was therefore natural to select the most critical component from a tribological viewpoint, namely the hydraulic pump, and develop suitable test procedures around various types of pump. Many different pump tests have emerged through the years, as technological developments progressed, and a selection of current interest are reviewed in Table 16.2.

16.2.2 Shear stability: IP 294, DIN 51 382, ASTM D3945, ASTM D2603 Hydraulic fluids intended for use over wide temperature ranges normally contain polymer additives to improve the viscosity index. When subjected to the high shear forces developed in certain parts of the system, e.g. in pressure-control valves and vane pumps, the polymeric viscosity index improvers are to a certain extent broken down mechanically. This results in a corresponding viscosity reduction, and the stability of a fluid against such shear losses is usually determined by circulating the fluid through a diesel injector nozzle, in accordance with IP 294, DIN 51 382 and ASTM D3945.

Samples are taken after the fluid has completed a predetermined number of passes (normally 250 cycles), and kinematic viscosities of the original fluid and of the sheared samples are determined at 40°C and 100°C. The percentage reduction in viscosity is thus a measure of the test fluid's shear stability as determined by the (Bosch) Diesel Injector Test Rig.

ASTM D2603 is an alternative test method utilizing sonic irradiation (10 kHz) to shear the polymer content of the fluid. However, it would appear that sonic oscillation rates the various types of polymer in a different order to mechanical devices, and correlation with pump tests and the diesel injector test is frequently poor.

Table 16.2 Test parameters for important hydraulic pump tests

Test	MPa	°C	rpm	h	Requirement
IP 281 – Vickers V-104C (vane pump)	14	≡13 cSt	1440	250	Total wear (ring + vanes) <100 mg
ASTM D2882 – Vickers V-104(5)C (vane pump)	13.8	80	1200	100	Total wear (ring + vanes) <50mg
Vickers 35VQ25A (vane pump)	21	93	2400	3×50	<75 mg (ring) <15 mg (vanes)
Denison P46 (axial piston pump)	35	57–85	2400	100	Minimal wear of phosphor-bronze slipper pads

17

CONTAMINATION

All hydraulic systems unavoidably contain varying amounts of con-taminants. A small 40-litre system, for example, might typically con-tain 1–3 000 000 particles smaller than 1 µm. More than 80% of all faults occurring in hydraulic systems are presumed to originate from contaminants in the functional fluid. It is therefore extremely impor-tant to control the contamination level in these systems, thereby reducing wear and extending component life.

17.1 What impurities are involved?

The impurities may be divided into two groups: (a) mechanical con-taminants (particles) and (b) chemical contaminants.

Mechanical contaminants (particles) have various origins and may be metallic, fibres, dust particles, etc. Generally speaking, it is the especially hard particles of a size somewhat greater than clearances between mating surfaces in relative motion that give rise to most wear in the equipment.

Chemical contaminants are predominantly moisture, degradation products from the fluid, cutting oils from machine tools, corrosion preventives and various cleaning fluids.

17.2 Where do the impurities originate?

Contaminants in a hydraulic system can be ascribed to the following origins:

'*Inherent particles*', such as foundry sand, oxide particles, dust and possibly other contaminants remaining from the installation period.

Generated particles, normally originating from wear processes affecting hydraulic components, but which may also be formed during assembly of the system.

The most important wear processes are:

- Adhesive wear ('scuffing'), caused by local cold welding of surface asperities on rubbing surfaces. On account of the relative motion, the welds are immediately torn apart. The metallic material is therefore transferred to and fro between the moving surfaces until, finally, wear particles loosen from the surfaces. Anti-wear additives hinder this wear mechanism, metal dithiophosphates and active sulphur–phosphorus compounds are particularly effective.
- Abrasive wear, caused by hard asperities or particles scratching or ploughing against a surface. Wear particles are generated by repeated plastic deformations, and abrasive wear is hence associated with a plastic fatigue mechanism.
- Surface fatigue, which first and foremost affects components operating under high contact pressures. According to Hertzian theory, the maximum shear force is developed just below the metal surface, and this can eventually initiate a fatigue crack. This crack propagates parallel to the surface for a short distance, then upwards, thus releasing a metallic particle (flake).

Even comparatively moderate moisture contents (<100 ppm) in hydraulic fluids can promote this destructive process. The small water molecules diffuse readily onto the reactive surfaces of newly-formed microcracks, decomposing to release atomic hydrogen. The hydrogen atoms then diffuse into the metal beyond the microcrack causing hydrogen embrittlement of the steel, thus assisting further propogation of the crack.

Fatigue particles generated in this manner will continue to promote further wear by generation of more particles.

External contaminants are impurities contaminating the hydraulic fluid via breather vents, access points, seals, leaking coolers, filling unclean fluid, and maintenance work.

Modern high pressure systems comprise a wide variety of sensitive components, requiring a high degree of cleanliness with respect to the hydraulic medium. Fig. 17.1 illustrates a number of the critical clearances found in typical components. Consequently, hydraulic systems are equipped with many different filters in order to protect the critical components and to maintain fluid cleanliness at an acceptable level.

Component Clearance (μm)

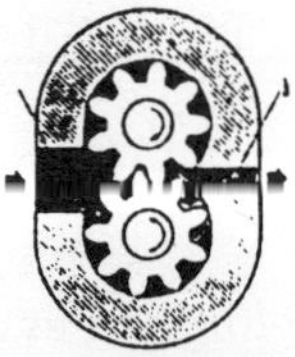

Gear Pumps

Gear to side plate	0.5–5
Gear tip to housing	0.5–5

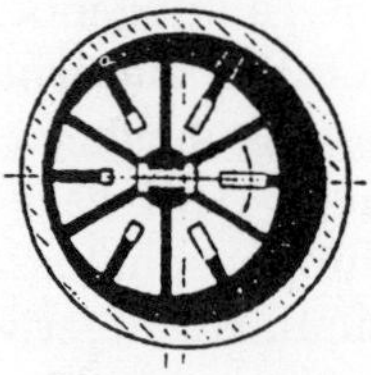

Vane pumps

Vane tip to stator	0.5–1
Vane to side plate	5–13

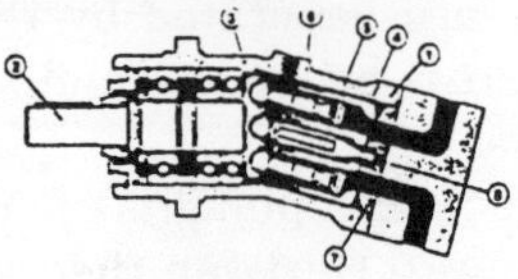

Piston pumps

Piston to cylinder	5–40
Cylinder to valve plate	0.5–5

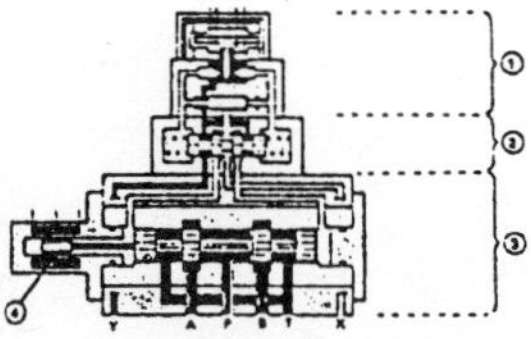

Servo valves

Jets	130–450
Spool to housing	1–4

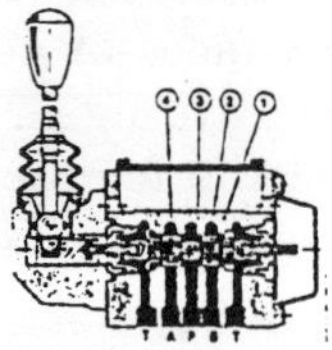

Control valves

Jets	130–10 000
Spool to housing	1–23
Seated valves	13–40

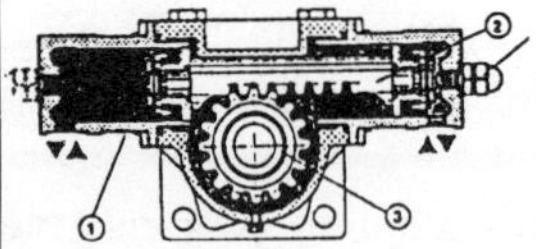

Actuators

Piston to cylinders	50–250
Hydrostatic bearings	0.5–25

Fig. 17.1 Typical dynamic clearances in hydraulic components.

The cleanliness level of a hydraulic fluid is normally specified in the form of a particle size distribution, against one of the standards mentioned below.

1. ISO 4406 (previously CETOP RP70), reporting particles of 5μm and 15 μm (Table 17.1).
2. NAS 1638, the American aerospace standard, reporting particles in five different size classifications (Table 17.2).

Table 17.1 Solid particle contamination level codes by ISO 4406

Code	No. of particles per ml > 5μm		No. of particles per ml > 15 μm	
	More than:	Up to and including:	More than:	Up to and including:
20/17	5000	10 000	640	1300
20/16	5000	10 000	320	640
20/15	5000	10 000	160	320
20/14	5000	10 000	80	160
19/16	2500	5000	320	640
19/15	2500	5000	160	320
19/14	2500	5000	80	160
19/13	2500	5000	40	80
18/15	1300	2500	160	320
18/14	1300	2500	80	160
18/13	1300	2500	40	80
18/12	1300	2500	20	40
17/14	640	1300	80	160
17/13	640	1300	40	80
17/12	640	1300	20	40
17/11	640	1300	10	20
16/13	320	640	40	80
16/12	320	640	20	40
16/11	320	640	10	20
16/10	320	640	5	10
15/12	160	320	20	40
15/11	160	320	10	20
15/10	160	320	5	10
15/9	160	320	2.5	5
14/11	80	160	10	20
14/10	80	160	5	10
14/9	80	160	2.5	5
14/8	80	160	1.3	2.5
13/10	40	80	5	10
13/9	40	80	2.5	5
13/8	40	80	1.3	2.5
12/9	20	40	2.5	5
12/8	20	40	1.3	2.5
11/8	10	20	1.3	2.5

Table 17.2 Cleanliness ratings by NAS 1638

Particle size (μm)	Class											
	00	0	1	2	3	4	5	6	7	8	9	10
						Number of particles per 100 ml						
5–15	125	250	500	1000	2000	4000	8000	16 000	32 000	64 000	128 000	256 000
15–25	22	44	89	178	356	712	1425	2850	5700	11 400	22 800	45 600
25–50	4	8	16	32	63	126	253	506	1012	2025	4050	8100
50–100	1	2	3	6	11	22	45	90	180	360	720	1440
>100	0	0	1	1	2	4	8	16	32	64	128	256

18

DETERIORATION AND MAINTENANCE

Reliability surveys reveal the majority of service problems are due to *contamination*, either in the form of particles promoting wear and blocking narrow clearances, or alien oil types (e.g. detergent engine oils) resulting in foaming and emulsion formation. A frequent source of particulate contamination is corrosion, induced by accumulation of excessive moisture in the system. Approximately 80–90% of all faults in hydraulic systems, abnormal wear, valve malfunctioning, etc., may be ascribed to particulate contamination and inadequate filtration of the hydraulic fluid (Table 18.1).

Finely divided contaminants also adversely affect demulsibility and air release properties of the fluid, and may catalyse the oxidation process. Conversely, only minimal wear and insignificant service problems are registered in hydraulic systems operating with a correct fluid type at high cleanliness levels. Cleanliness is therefore of prime importance and is achieved by thorough flushing of new systems, charging with microfiltered (5 μm) fluid, and use of suitable filters during operation.

In critical systems particle counts should be considered, these providing a useful guide as to whether the filtration level selected for the system is satisfactory for operational reliability and durability of system components. Figure 18.1 indicates the reationship between cleanliness level and operational pressure to achieve acceptable component life. Regular particle counts will reveal whether the filters are functioning satisfactorily, the ingress of extraneous contaminants, whether mechanical contamination is increasing, and provide early warning of potential service problems. Particle counts may be supple-

Table 18.1 Suggested cleanliness levels for various types of system.

Desirable cleanliness level by ISO 4406		Maximum number of particles per 100 ml of size		Sensitivity	Type of system	Recommended degree of filtration (µm) (βx > 100)*
5 µm	15 µm	5 µm	15 µm			
13	9	8 000	500	Super critical	Silt-sensitive control system with very high reliability. Laboratory or aerospace	1–2
15	11	32 000	2 000	Critical	High-performance servo and high-pressure long-life systems, i.e. aircraft, NC machine tools, etc	3–6
16	13	64 000	8 000	Very important	High-quality reliable systems. General machine requirements	6–10
18	14	250 000	16 000	Important	General machinery and mobile systems. Medium pressure, medium capacity	10–15
19	15	500 000	32 000	Average	Low-pressure, heavy industrial systems or applications where long life is not critical	16–25
21	17	2 000 000	130 000	Highly tolerant	Low-pressure systems with large clearances	25–40

* βx = 'beta factor' of a filter. It is equivalent to the number of particles of size x retained by the filter divided by the number of similar particles passing through the filter, when evaluated by ISO–4572.

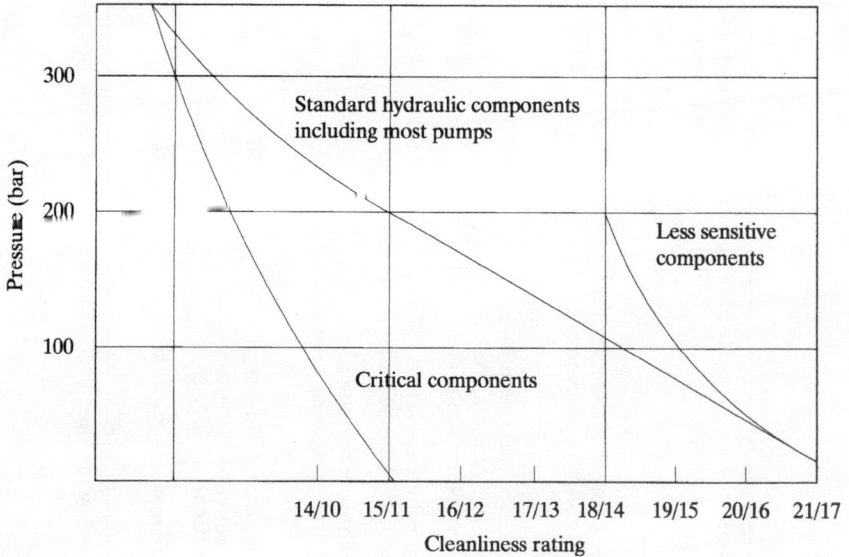

Fig. 18.1 Suggested cleanliness ratings for acceptable component life (assuming operation within recommended viscosity range).

mented by spectral analyses identifying the exact types of wear metal present (iron, copper, lead, nickel, etc.), and/or ferrography providing additional information on the origin of the particles.

Oil level, service temperature and filters should be checked at regular intervals. It is also important to drain possible separated water and sedimented contaminants from a drainage point located at the lowest point of the oil tank. 'Breathing' by the fluid reservoir may, over a period of time, result in the accumulation of considerable amounts of condensation water, especially in systems operating at moderate service temperatures and/or in humid environments.

Water in a hydraulic oil increases the danger of corrosion and wear in valves and the hydraulic pump. In outdoor units, the viscosity increase due to emulsified water can be very significant when attempting to start the system at low ambient temperatures, and valves may be inoperative on account of ice formation.

A mineral oil-based hydraulic fluid will normally exhibit a cloudy (milky) appearance when the content of dispersed water exceeds 0.05% (500 ppm), and in most industrial systems this should not be allowed to rise above 0.10–0.15% . The principal exceptions are various robust, low pressure units (operating pressures ≈3.5 MPa) without sensitive components, in which a water content of 0.25–0.40% may be acceptable. At the other extreme are certain critical high pressure systems for which manufacturers specify moisture limits of 200 ppm or less.

When the water content of a hydraulic oil exceeds the recommended value, the oil should be replaced by a fresh charge, although in the case of large oil volumes it may be economic to gain a temporary respite by centrifuging or settling the oil. Another possible alternative treatment to remove relatively small volumes of water is the use of water absorbing filter elements, now available from a number of suppliers.

The oil will also be subject to a continuous *chemical* degradation, principally oxidation, though apart from a certain darkening in colour, the oil's properties will hardly be affected during the initial period of use. Later, the rate of oxidation will accelerate, resulting in higher acidity, increased viscosity, decreased water separation and a tendency to form resinous deposits. On account of this latter lacquering tendency, it is unwise to delay too long in replacing an oxidized oil, particularly in a hydraulic system incorporating sensitive servo-valves.

At operational temperatures up to 70°C, the oxidative life of a mineral oil-based hydraulic fluid of reasonable quality will normally be several thousand hours of service. At higher temperatures the oil life may be appreciably shorter as the rate of oxidation increases exponentially with rising temperature (see Chapter 9).

Electrical heating elements, sometimes incorporated to facilitate easy circulation during the initial start-up period, are best avoided as they are frequently over-dimensioned and cause local overheating with thermal degradation of the fluid. When used, the *maximum* recommended power density is 0.8 W/cm^2 surface area.

The degree of oxidation, i.e. the level of chemical deterioration, may be measured by means of several parameters, for example an acidity determination (TAN). The increase in acidity with time is, however, closely related to the composition of the oil, and it is not possible to prescribe any general limiting value. As is shown in Fig. 9.6 (see Chapter 9), oils containing metal dithiophosphate additives are characterized by a relatively high initial acidity. This acidity stems from the amphoteric nature of the additive and decreases during the initial period of service, subsequently rising as the oil oxidizes. In contrast, oils inhibited with phenolic antioxidants display a relatively low and stable acid value over a long period of service, until the induction period is passed. Interpretation of acidity levels should therefore be left to the oil supplier unless the composition of the product is known or sufficient experience has been gained with the oil quality in question.

Polymer-containing hydraulic oils of particularly high viscosity index become progressively less viscous during use due to mechanical degradation of the polymeric viscosity index improver under high

	One drop per second	Drip over to trickle	3 mm jet
Litres/hour	0.3	4	41
Litres/24 hours	7.0	90	1000
Litres/month	200	2700	31000

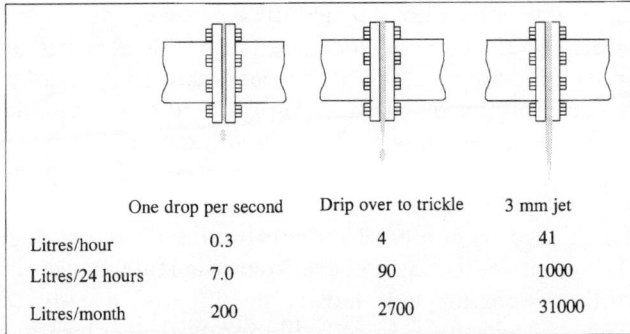

Fig. 18.2 Cumulative loss of hydraulic fluid by leakages.

shear stresses. A common recommendation is to consider replacing the oil charge when the viscosity loss is equivalent to 10–15% of the oil's *measured* viscosity when new (not the nominal ISO VG mean value).

Correctly formulated mineral oil-based hydraulic oils are unharmed by frost, but should preferably be stored under dry conditions, or in any event under cover when stored outdoors. In general, and *always* in the event of uncovered outdoor storage, the oil drums should lie on their sides with the tapping plugs at '03:15 hours', and on some form of support so as to avoid direct contact with the ground. The reason for this is to avoid rainwater and other contaminants accumulating on top of the drums, and being gradually drawn in via the tapping plug threads as the drums 'breathe' with temperature variations.

Even apparently insignificant leakages can result in surprisingly large losses of hydraulic fluid over a period of time (Fig. 18.2). Leakages not only represent financial loss, but also poorer working conditions and safety hazards in the form of slippery floors and fire risk.

18.1 Flushing

Flushing may often be required when the system is new or during subsequent service, particularly after repairs, if significant contamination has ocurred. Large systems assembled on site frequently incorporate components treated with temporary corrosion protectives. Unless such materials are removed prior to filling the system with a clean hydraulic fluid, they will gradually dissolve in the working fluid and may detract from its performance in several respects, e.g. by promoting emulsification.

If flushing is required, it is preferable to utilize the grade to be used in service or a lower viscosity grade of similar composition. Critical components (valves, hydraulic pumps and motors) should preferably be isolated or by-passed during the flushing operation to avoid harmful accumulations of particulate contaminants, thus defeating the purpose of the treatment. A minimum charge of flushing oil should suffice, this being circulated through the system and adequately dimensioned supplementary filter(s) by a suitable external pump. Filters should be checked periodically and flushing continued until particle counts on the fluid returning to the filters decrease to a satisfactory level.

The temperature of circulation should preferably be around 40°C; if necessary the flushing medium should be heated, but local overheating must be avoided. If electrical heaters are used they should be of liberal surface area. As a guideline, trials with a typical unit revealed a surface temperature rise of 40°C with a power density $\approx$0.8 W/cm^2, and 60°C using $\approx$1.6 W/cm^2. When a high pressure pump is available for the purpose, the temperature of the flushing medium may be conveniently raised by passing it through some form of choke, e.g. a pressure relief valve or a length of narrow bore piping immersed in the fluid reservoir.

19

ANALYSIS OF USED HYDRAULIC OIL

The condition of a hydraulic fluid is normally monitored by periodical determination of viscosity, total acidity, water content and visual inspection with respect to mechanical impurities. On the basis of these results, it can be decided whether the oil is suitable for further use. A more complete assessment of the oil condition may be obtained by supplementing the above-mentioned results with one or more of the following tests:

- *spectral analysis* (AA or ICP), for identification of wear metals
- *particle counts*, for evaluation of mechanical impurities, numerically and by size distribution
- *ferrography*, to determine the origin and morphology of the metallic particles
- *infrared absorption analysis* (IR), to examine the chemical composition of the oil and degradation products
- *oxidation tests*, for estimating the remaining useful life of the oil

The application of these test methods yields a significantly more reliable assessment of the oil condition, but are relatively expensive to perform and hence are utilized with discrimination. Oxidation tests can be particularly time-consuming and are therefore rarely applied during routine condition monitoring.

Table 19.1 lists the test results of samples drawn from a mobile hydraulic system after 250, 1500 and 3000 h of operation. The oil type is an ISO HV/VG32, the unused oil having a viscosity index of approximately 180, and meeting the performance requirements of DIN 51 524 (H-LP).

Table 19.1 Analytical results for a hydraulic oil (ISO type HV, VG32) after 250, 1500 and 3000 h of use

Analysis	New oil	Used 250 h	Used 1500 h	Used 3000 h
Kinematic viscosity cSt at				
100°C	7.0	6.65	6.3	5.9
40°C	32	30.8	29.8	27.9
Flash point (°C)	165	170	168	163
Water content (ppm)	40	180	1100	1800
TAN (mg KOH/g)	0.75	0.52	0.65	0.87
Additive metals				
Zn (%m)	0.04	0.04	0.03	0.02
P (%m)	0.04	0.04	0.03	0.03
Appearance/colour	Clear, pale yellow	Clear, pale yellow	Slightly cloudy, dark yellow	Cloudy, amber
Wear metals, ppm				
Iron (Fe)	1	6	29	53
Copper (Cu)	0	1	5	11
Chromium (Cr)	0	< 1	1	2
Lead (Pb)	0	< 1	1	3
Aluminium (Al)	0	< 1	1	2
Tin (Sn)	0	< 1	2	3
Silicon (Si)	< 1	1	4	8

19.1 Interpretation of the test results

From the analyses in Table 19.1 we may conclude that the zinc dithiophosphate-based hydraulic oil shows only minor changes after 250 h, is still in reasonably good condition after 1500 h, and nears the end of its serviceable life at 3000 h of operation on account of oxidation, viscosity change and increasing contamination.

Viscosity has decreased progressively due to mechanical shearing of the polymeric viscosity index improving additive, this phenomenon overshadowing any possible viscosity increase due to oxidation.

Flash point increases slightly during the initial period of operation. This is ascribed to gradual evaporation of the most volatile hydrocarbon components. Subsequent minor changes reflect the combined effect of evaporation and formation of volatile oxidation products.

Water content rose immediately on filling the system due to residues of the previously (used) oil. The water content increased to 1100 ppm (0.11%) after 1500 h, and further to 1800 ppm (0.18%) after 3000 h of operation. These increases are principally due to condensation moisture, as the oil's oxidative deterioration is not yet far enough advanced to make any significant contribution to the measured water content.

Total acidity (TAN) for the unused oil originates from the amphoteric nature of the dithiophosphate additive, and the initial value is still falling after 250 h as the dithiophosphate is activated. The acidity probably falls to a value of 0.15–0.20 during further operation, then rises again with the onset of hydrocarbon oxidation reactions. The total acidity is still at a satisfactorily moderate level after 1500 h, but later exceeds the original value of the new oil. After 3000 h the acidity has risen to 0.87, indicating that the oxidation rate is increasing. There may now be a risk of forming sticky, resinous deposits of polymeric oxidation products, causing malfunctioning of critical servo-valves, etc.

Additive content is monitored by measuring the concentrations of zinc (Zn) and phosphorus (P) remaining in the oil phase. The concentrations of these elements are gradually reduced as the dithiophosphate additive undergoes chemical reaction and is progressively depleted, mainly by film-forming reactions at metal surfaces. This eventually results in reduced anti-wear properties of the oil in circulation.

Appearance and *colour* still resemble those of new oil after 250 h, but the oil becomes progressively darker and cloudier due to chemical deterioration and increased content of dispersed water.

Wear metals contaminate the oil to varying degrees, depending upon the metallurgy of the components subject to wear or corrosion. In the present case, silicon (Si), copper (Cu), lead (Pb) and tin (Sn) are the elements of principal interest.

The rising values of Pb, Sn and Cu derive from the combined influence of acidic oxidation products, the active dithiophosphate additive, and general abrasive wear.

The silicon content of the unused oil is due essentially to the anti-foam additive, a silicone oil. The subsequent increase during service is mainly the result of contamination (silicaceous dust), this promoting abrasive wear throughout the system.

Ferrous metals predominate in components and the system in general. Not surprisingly, iron (Fe) is present at a relatively high concentration in the used oil; the measured concentration is, however, considered reasonably normal for the actual hours of operation.

Determination of the wear metals present in a used oil sample may therefore yield valuable information with respect to the wear pattern of a particular system under the actual conditions of operation.

19.2 Condition monitoring and oil change

It is hardly remunerative to perform regular *laboratory analyses* on small systems ($\leq$ 250 litres), and oil change intervals are usually decided on the basis of experience. Large hydraulic systems, however, often contain several tons of hydraulic fluid, and the condition of the oil charge is normally checked by *regular analyses* (e.g every sixth month) (see Table 19.2).

Table 19.2 Oil change intervals

Small systems ($\leq$ 250 litres)	Large systems
According to experience, e.g.:	Determined by:
A) At operating temperatures 40–50°C: 1.5–3 years	A) Laboratory analyses every 6–12 months
B) At operating temperatures 60°C: 1 year	B) Regular visual examination

As a safety measure against accidental contamination and unexpectedly rapid deterioration, it is prudent to make weekly *visual examinations* of hydraulic fluids in service as shown in Table 19.3.

Table 19.3 Condition monitoring by visual inspection

Appearance Immediate	After 1 h	Cause	Action
Clear, normal odour	—	Normal	None
Cloudy	Clear	Air entrainment	Check system
Cloudy	Clear, free water	Unstable emulsion	Drain off water regularly
Cloudy	Clear	Stable emulsion	Analyse oil, consider oil change
Dirty, normal odour	Sediment	Contaminated	Analyse oil, check filters
Very dark, sharp odour	Unchanged	Oxidized	Sample to laboratory

20

FIRE-RESISTANT FLUIDS

The most common types of fire-resistant medium are:

- oil-in-water (o/w) emulsions, classified as ISO-type HFAE,
- water-in-oil (w/o) emulsions, ISO-type HFB,
- water–glycol solutions, ISO-type HFC,
- phosphate esters, ISO-type HFDR,
- organic esters, ISO-type HFDU.

Although these products present a significantly lower fire hazard than conventional mineral oil-based fluids, the individual fire-resistant products often possess certain other less advantageous properties; examples are given below.

(a) The relatively high density of some fire-resistant fluids results in greater demands on the pump's suction capacity in systems where the pump is positioned above the fluid level.

(b) Certain fluids with low viscosity indices are considerably more viscous than conventional hydraulic fluids at low ambient temperatures. In such cases positioning pumps below the fluid reservoir in order to obtain a positive suction head may be necessary to avoid cavitation problems prior to achieving the normal operational temperature.

(c) The high vapour pressure of aqueous fluids may easily induce bubble formation at low pressures and cause cavitation. The temperature of operation for ISO types HFAS, HFAE, HFB and HFC should therefore be limited to 55–65°C to obviate this problem and restrict evaporation losses from the water phase.

Eventual evaporation losses should be compensated for by regular additions of demineralized water; this is particularly important with respect to HFC fluids on account of the pronounced thickening tendency as the water content decreases.

(d) Microbiological growth is a potential danger in most systems utilizing water-based fluids, especially in HFAE and HFAS media at moderate temperatures of operation.

(e) The lubricating properties of aqueous media are generally inferior to those of conventional mineral hydraulic fluids, and this shortcoming may often necessitate modifications to the system when converting from mineral fluids to aqueous hydraulic fluids. The fatigue life of ball and roller bearings is considerably reduced when lubricated by fluids containing water; special pump designs incorporate over-dimensioned bearings to compensate for this shortcoming.

(f) A number of fire-resistant fluids attack conventional seal materials, hoses and paints. This particularly concerns phosphate esters, water–glycol solutions and certain organic esters.

The properties of various fire-resistant fluids are compared in Table 20.1.

HFAE media are oil-in-water (o/w) emulsions, containing by definition a maximum of 20% mineral oil finely dispersed within the water phase. Consequently, the viscosity of these fluids approximates to that of water ($\approx$0.8 cSt at 40°C), and conventional hydraulic systems intended for use with a mineral hydraulic fluid normally require modification in order to utilize the HFAE fluid without causing excessive wear of critical components. Operational limitations for these fluids are pressures $\leq$6 MPa and a temperature range of 5–50°C. Nevertheless, o/w emulsions have given satisfactory service for many years in large systems operating at low pressures. An inherent property of aqueous media is the unfortunate reduction in fatigue life of heavily loaded steel surfaces. It is therefore preferable to utilize journal bearings rather than ball or roller bearings in hydraulic pumps and motors operating on these fluids. The quality of water used in preparing these emulsions is also extremely important; hard-water qualities may result in calciferous deposits, whilst soft water promotes foaming. Biocide additives are frequently incorporated in aqueous media to prevent microbiological growth and associated problems, always a potential hazard in these fluids at moderate operating temperatures. The alkaline nature (high pH) of many formulations represents a potential corrosion hazard with respect to magnesium, zinc, cadmium, lead and aluminium. Recent development work has resulted in the formulation of even finer o/w dispersions ('microemulsions')

Table 20.1 Comparison of various fire-resistant fluid media.

ISO-type:	Mineral oil HM	Phosphate ester HFDR	Water/glycol HFC	Organic ester HFDU	O/W emulsion HFAE
Density at 15°C	0.88	1.15	1.10	0.93	1.0
Viscosity index	High	Low	Very high	Very high	High
Vapour pressure	Low	Low	High	Low	Very high
Lubricity	Good	Good	Moderate	Good	Limited
Roller bearing load factor	1	1.2	2.6	1	High
Metals to be avoided	—	Aluminium (sliding contacts)	Zinc, cadmium, magnesium	—	Depends on pH value
Recommended paints	Epoxy	Epoxy	Epoxy	Epoxy	Epoxy
Seal materials, O-rings*	Nitrile, Viton, polyurethane	Butyl, EPR, Viton	Nitrile, Viton	Nitrile, Viton	Nitrile, Viton
Hose materials, (external: neoprene)	Nitrile	Butyl, EPR, Viton	Nitrile	Nitrile, Viton	Nitrile
Max. operating temperature (°C)	130	150	65	150	50
Bulk modulus (10^9 N/m^2)	1.5	2.0	2.9	2.2	High
Fire resistance	Poor	Excellent	Good	Good	Excellent
Relative price (mineral oil = 1)	1	5	3	4	0.2

* On account of the complex composition of proprietary materials, the behaviour of seals based on nominally identical elastomers may vary. Screening by immersion tests on the specific products at realistic temperatures is recommended when new materials are introduced.

possessing improved stability towards separation of the oil phase, together with superior lubricating properties. Modern products contain typically 3–10% mineral oil and are popularly known as *HWBF-fluids* ('high water base fluids'). The search for further improvements in lubricating properties has resulted in development of HWBF products of higher viscosity, e.g. Shell Irus Fluid AT with a kinematic viscosity of 11 cSt at 40°C. Such fluids can probably replace conventional mineral hydraulic oils in many applications without any particular modification of the systems, and promising results have been achieved, e.g. in the steel industry (CON-CAST units).

HFB fluids are water-in-oil (w/o) 'invert' emulsions consisting of minute water droplets, 1–2 μm in diameter, finely dispersed within the oil phase. The water content is normally around 40%v in order to confer satisfactory fire resistance and meet the relevant specification requirements. Marketed products are usually ISO VG68 or 100, and may be used over a temperature range of 0–50°C, or even up to ≈65°C if the increased volatility is acceptable. HFB fluids naturally possess far better lubricating properties than HFAE media, and yield satisfactory performance in gear, screw and axial piston pumps up to 20 MPa. Vane pumps, however, rate the lubrication properties of HFB emulsions more critically, and operating pressures above 5–6 MPa are not generally recommended.

HFC fluids are aqueous solutions of glycols and polyalkylene glycols, with the addition of various additives, e.g. anti-wear, corrosion and foam inhibitors. The water content should be at least 35% by volume to ensure satisfactory fire-resistance properties (the majority of products marketed contain 40–45%), and this should be monitored regularly and maintained at the specified level to obviate troublesome viscosity increases. An interesting, though not generally recognized, fact is the temperature susceptibility of the lubricating properties of some HFC fluids. With certain formulations, for example, the anti-wear performance may be significantly improved by increasing the operational temperature towards the upper limiting value of 65°C. Systems intended for HFC fluids should not incorporate components of magnesium, cadmium or zinc. The fluids are compatible with conventional nitrile rubber seals, but are extremely aggressive to many types of paint coating. HFC fluids normally tend to entrain air and hold particulate contamination in suspension to a greater degree than mineral hydraulic media. Consequently, these fluids are more critical with respect to effective de-aeration by good reservoir design, and adequate filtration capacity.

HFD fluids are formulated on the basis of various synthetics, and may be classified according to their chemical composition. The base fluids include organic esters of diverse complexity, polysiloxanes, silicate esters, phosphate esters and polyphenyl ethers. Synthetic hydrocarbons derived from polymerization of α-olefin oligomers have found extensive use in the formulation of less flammable hydraulic fluids for the aviation industry (see Chapter 14).

A number of the HFD-fluid base materials have become considerably less attractive during recent years, due to the increased focusing on environmental factors. This especially concerns HFDS and HFDT fluids on account of their halogen contents. However, the unique properties associated with certain of these halogenated fluids will undoubtedly ensure continued interest in further development for aerospace and military applications.

Among HFD media it is principally the phosphate esters and organic esters that are selected for industrial applications today. Of these, the phosphate esters are clearly the most fire-resistant, although they inevitably present a number of limitations with respect to the design of the system and choice of materials (Table 20.1). The high density, low viscosity indices and poor low temperature flow properties of many phosphate esters necessitate due attention to the dimensioning and positioning of suction lines, filters and strainers. Extreme caution must be exercised in the selection of seals and hoses. The incompatibility of these fluids with conventional nitrile elastomers frequently precludes conversion of mineral-oil systems due to the prohibitive cost of replacing all seals, etc., by fluorocarbon elastomers, EPR or butyl rubber alternatives. As in the case of all fire-resistant fluids, it is wise to confer with suppliers of both seals and components to ensure satisfactory compatibility with the selected hydraulic medium.

20.1 Conversion of existing systems to fire-resistant fluids

Unless specifically designed for a particular fire-resistant medium, most hydraulic equipment is normally designed for the use of mineral hydraulic oils. Successful conversion to fire-resistant fluids is dependent upon careful selection of fluid types suitable for the system components (with eventual modifications) and operational parameters. In addition, satisfactory condition monitoring, contamination control and maintenance routines must be established.

When converting existing systems, the following guidelines are suggested:

1. Obtain comprehensive technical information regarding the system components, conditions of operation and manufacturer's

recommendations. The component suppliers (particularly the pump manufacturers) often issue their own requirements for use of fire-resistant fluids, frequently issuing lists of approved products and recommending necessary system modications.

2. On the above basis, discuss with the suppliers which fluid types could be suitable, and whether modifications to the system are required. Possible modifications might include pressure and/or rpm limitations in systems employing water-based fluids. Similarly, it could be advisable to utilize different filter elements, and replace light alloy components by stainless steel alternatives.

3. Drain the previous fluid completely from the system, dismantling components, filters, couplings, etc., as required to empty the system.

4. Wipe the internal walls of the fluid reservoir clean and dry.

5. Clean and change all filters. Check that the type of filter is suitable, i.e. compatible and of acceptable efficiency and capacity for the selected fluid.

6. Remove eventual paint and tank coatings with caustic soda or other proprietary paint stripper should there be the slightest doubt regarding their chemical resistance to the new fluid. Be aware that external painted surfaces will probably also be subject to a certain degree of contact with the working fluid on account of possible spillages or leakages. *(N.B. All pipe connections should be securely sealed off so as to protect the rest of the system during this operation.)*

7. Replace all incompatible components, seals and flexible hoses.

8. Check that the suction line and strainer are adequately dimensioned with respect to the flow and volatility characteristics of the new fluid.

9. If the system contains significant residues of mineral oil, fill with a minimum charge of fire-resistant fluid and circulate for 30 min. Thereafter drain the system and fill with new fluid.

10. Inspect all filters and strainers at regular intervals during the initial period of operation. Monitor the condition of the new fluid by relatively frequent laboratory checks until reassured that the system is functioning satisfactorily.

20.2 Maintenance of fire-resistant fluids

The user of fire-resistant fluids must accept the need for greater vigilance with repect to fluid maintenance than is normal for conventional mineral hydraulic fluids. No longer is it sufficient to periodically

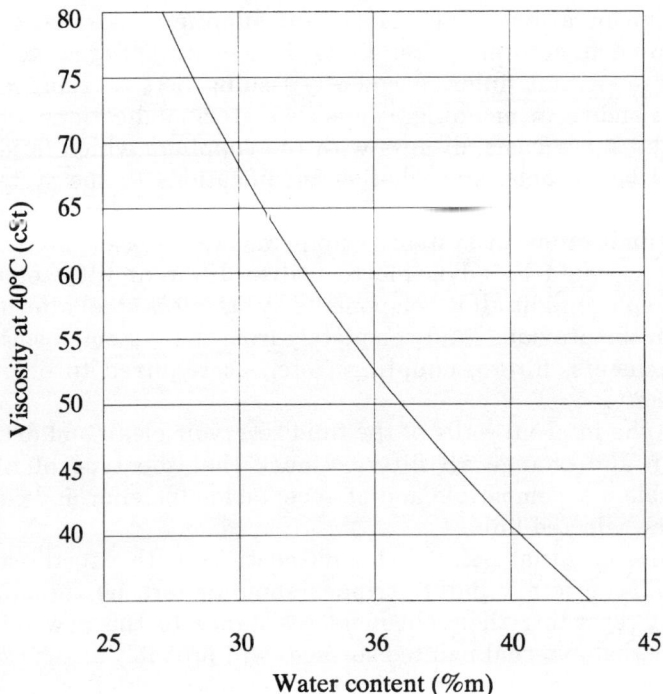

Fig. 20.1 Viscosity–water content relationship for ISO HFC fluid (Shell Irus Fluid C).

change filters, repair the occasional leak (fire-resistant fluids are expensive!) and top up the fluid reservoir.

The majority of fire-resistant fluids display a considerably greater degree of detergency than conventional mineral hydraulic fluids, and consequently dirt particles or wear debris do not settle out in the reservoir as readily. This tendency promotes abrasive wear of pumps and other components besides causing valves to malfunction.

Effective contamination control is therefore an important aspect of operation with fire-resistant fluids and a simple settling test, performed by allowing a representative sample of the fluid in circulation to stand for 24 h, can yield a useful indication of filtration efficiency.

Evaporation losses from invert emulsions (HFB fluids) and water-glycols (HFC fluids) may have unfortunate secondary effects. Reduced water content in HFB fluids results in decreased fire resistance and viscosity, whilst water–glycol solutions become more viscous (Fig. 20.1). The viscosity of HFC fluids varies inversely with the water content and is often used as a simple means of assessing evaporation losses.

Table 20.2 Monitoring programme for fire-resistant fluids ISO-type HFAE, HFB, HFC and HFDR

O/W emulsion (HFAE)	W/O (HFB)	Water–glycol (HFC)	Phosphate ester (HFDR)
Oil content	Water content	Water content	Density
Emulsion stability	Emulsion stability	pH	Acid value
pH	Viscosity	Viscosity	Viscosity
Microorganisms	Microorganisms	Microorganisms	Moisture
Particle count	Particle count	Particle count	Particle count

Once a correct combination of system design and hydraulic fluid is established, the key to economic and effective operation is strict adherence to manufacturers' recommendations, systematic inspection of filters, and periodical monitoring of the hydraulic fluid by laboratory examination as indicated below in Table 20.2. When laboratory analysis confirms the condition of the fluid is no longer acceptable, the system should be drained and re-charged with fresh fluid. Prolonged use of a fluid that has deteriorated beyond accepted quality limits is faulty economy and will eventually result in costly down time and component expenditure.

21

HYDRAULIC BRAKE FLUIDS

Since the introduction of hydraulic brake systems during the 1920s, the development of synthetic elastomers and availability of new chemical intermediates, on-going research has resulted in improved brake-fluid quality to keep pace with automotive developments.

The essential function of a brake system is to convert kinetic energy of the moving vehicle into heat. Much of this heat is dissipated to the surrounding air, but a considerable amount of heat energy is inevitably absorbed by the brake fluid. The braking systems of modern high speed cars, heavy commercial vehicles, etc., are subject to increasingly high temperatures, and the brake fluids must therefore possess adequate thermal stability and low vapour pressure. In addition the fluid must display suitable flow properties over the entire range of anticipated operational temperatures, lubricate, protect all metal components against corrosion and be fully compatible with seals and elastomers.

Automotive brake fluids are formulated to satisfy the requirements of important international specifications. Most familiar are SAE (Society of Automotive Engineers) Standard J1703f and Federal Motor Vehicle Standard No. 116 (FMVSS116). The latter Federal standard classifies brake fluids in three categories of increasing severity, in accordance with Department of Transportation DOT3, DOT4 and DOT5. The salient features of the three classes are shown in Table 21.1.

Apart from fluidity considerations, interest is principally centred on achieving stable high boiling points, in order to avoid decreased braking power due to excessive volatility. A major failing of early

Table 21.1 Comparison of essential requirements for DOT3, DOT4 and DOT 5 brake fluids

	DOT3	DOT4	DOT5
Kinematic viscosity (mm^2/s) at $-40°C$, (minimum)	1500	1800	900
Minimum dry equilibrum reflux boiling point (°C)	205	230	260
Minimum wet equilibrum reflux boiling point (°C)	140	155	180

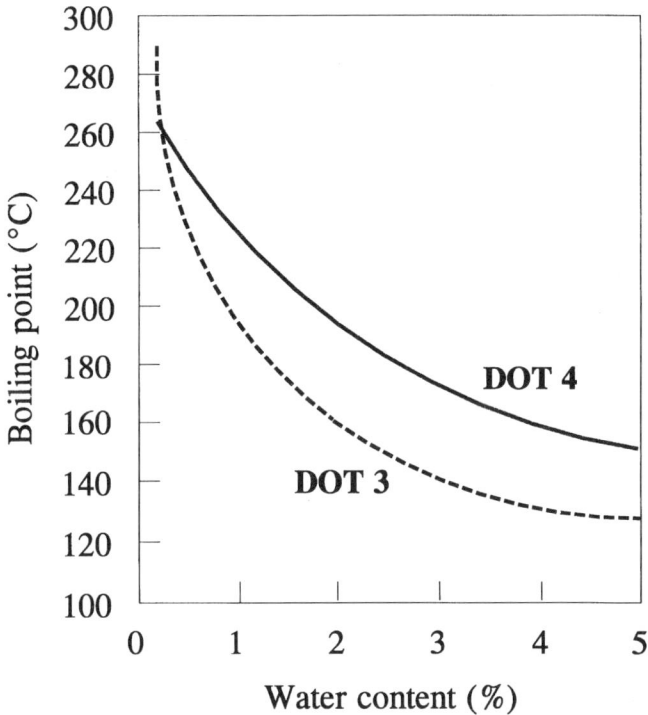

Fig. 21.1 Effect of water absorption on boiling point of FMVSS 116 DOT3 and DOT4 brake fluids (typical).

DOT 3 fluids, based on polyglycols and glycol ethers, was their inherent hydroscopicity and subsequent deteriorating boiling point as moisture was absorbed (Fig. 21.1). Surprisingly large amounts of moisture may accumulate in these fluids; under normal(!) UK climatic conditions absorption figures of around 2% per annum have been registered. Understandably, the progressive accumulation of moisture causes dramatic changes in the overall performance of a brake fluid:

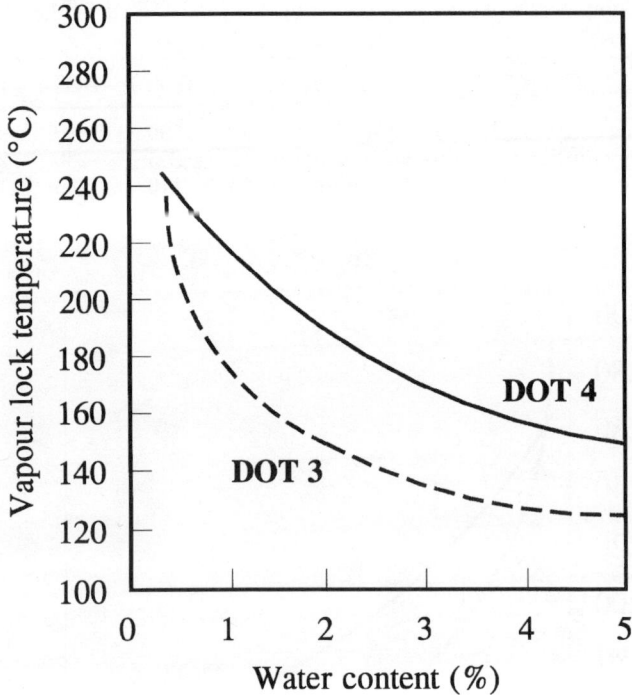

Fig. 21.2 Effect of water absorption on vapour lock temperature of FMVSS 116 DOT3 and DOT4 brake fluids (typical).

- *Increased low temperature viscosity,*
- *Reduced vapour lock value,*
- *Reduced anti-corrosion properties.*

The deleterious effect of water contamination is alleviated in DOT4 brake fluids by incorporation of borate esters in the final formulation. These components react chemically with absorbed moisture to form glycol ethers and thereby delay the onset of any vapour lock tendency or undue viscosity increase (Figs 21.2 and 21.3).

A further asset is the boosted corrosion protection afforded by the borate ester. By means of borate ester technology it has even proved possible to comply with the basic requirements of DOT5.

DOT5 fluids are, however, normally based on silicone oils (poly-siloxanes) and silicate esters.

Polysiloxanes possess excellent viscosity–temperature characteristics, are not hygroscopic and sustain a high boiling point over an indefinite period. These silicone base fluids are, however, immiscible with both water and the other DOT categories. Although immiscibility

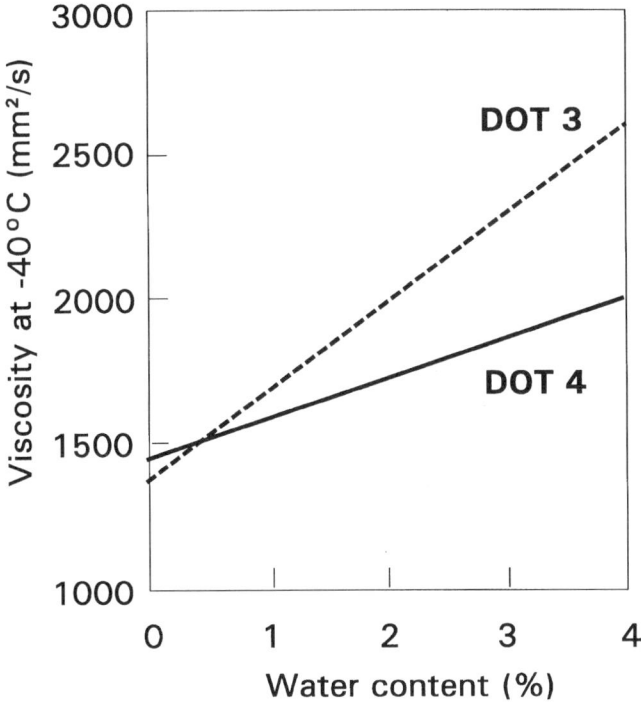

Fig. 21.3 Effect of water absorption on viscosity at −40°C for FMVSS 116 DOT3 and DOT4 brake fluids (typical).

with water ensures that the high boiling point does not decrease, there is a need for other components to eliminate any risk of incidental free moisture freezing inside the brake system. Nor are the polysiloxane base fluids fully compatible with the standard elastomers utilized in modern brake systems, requiring the addition of suitable seal swell additives to counteract shrinkage of the standard seal materials.

Silicate esters also exhibit excellent viscosity–temperature properties and have the advantage of good miscibility with current glycol ether-based brake fluids. In many conventional hydraulic applications the inherent hydrolytic instability of silicate esters has been regarded as a shortcoming. In automotive brake systems, where a certain ingress of moisture is inevitable, the ability of these fluids to absorb and chemically combine with the moisture is a valuable asset. This mechanism resembles the action of the added borate esters in DOT4 media, and hinders deterioration of the inherent favourable boiling point.

Proprietary brake fluids complying with DOT5 are popularly termed 'silicone' fluids, but are blended products incorporating suitable concentrations of other components to supplement and modify the properties of the base fluid.

Hydrocarbon base fluids resemble silicones in their lack of hygroscopy and deterioration of boiling point is therefore not a problem with these media. Hydrocarbon fluids normally require the addition of viscosity index improvers to approach the excellent viscosity indices of silicones, although some synthetics, e.g. polyalphaolefins, possess inherently high indices and also impressive low temperature flow properties. The lubricating properties of hydrocarbons are generally superior to silicones, but they too are not fully compatible with certain elastomers commonly used in automotive brake systems, namely ethylene–propylene rubber (EPR) and styrene–butadiene (SBR). Apart from the question of seals, the main objection to hydrocarbon fluids is their hydrophobic nature and risk of moisture freezing within the brake system. To alleviate this potential hazard, it is usual to incororate some form of water-scavenging agent in the final formulation, as in the case of silicone-based fluids. The natural ability of hydrocarbons to protect metal components against corrosion is a distinct advantage in brake systems, where a certain ingress of moisture is to be expected, and hydrocarbons are superior to silicones and glycol ether fluids in this respect.

Despite the apparent suitability of hydrocarbon fluids as base materials, the fear of accidental contamination by other mineral oils has been a distinct constraint to their widespread application. Nevertheless, manufacturers such as Citröen (since 1967) and Rolls-Royce successfully use hydrocarbon-based fluids in their central hydraulic systems. Hydrocarbon fluids are still specified for certain military applications, although there is a general trend towards specialized fire-resistant media for military and aviation systems.

Most agricultural machines, tractors, etc., utilize mineral oil products in their brake systems, even though some braking systems employ separate hydraulic circuits independent of the fluid transmission. Most tractor brake systems are, however, integrated with the main transmission system, thus imposing further quality constraints upon the transmission fluid formulation.

Modern tractors are usually equipped with oil-immersed ('wet') brakes which complain audibly with the characteristic 'wet-brake squawk' when unsuitable oil formulations are employed. Wet-brake squawk is an audible manifestation of stick–slip conditions, emanating from an unsuitable combination of kinetic and static friction coefficients for the lubricated surfaces of the brake friction discs.

Suitable fluids are formulated using carefully selected friction modifiers which are adsorbed onto the friction surfaces, minimizing the difference between the static and kinetic frictional forces, without unacceptably lowering the torque developed by the brake. Friction modification smooths the engagement of the friction discs by minimizing stick–slip, at the same time maintaining torque levels within design parameters.

Insufficient friction modification results in squawk and wear of the discs, whilst too low friction permits slipping and glazing of the mating surfaces due to oxidation of the transmission oil. In order to accommodate this requirement, and that of immersed multiplate clutches serving the power shift transmission, independent power take-off and power shuttle reverse boxes, the formulation of 'universal tractor transmission oils' (UTTO) requires a delicate balance of anti-wear, detergent, friction modifiers and other additives.

Wet brakes were introduced in Europe some few years after their appearance in the United States. During this interval in the 1960s, the ingenious idea of a truly universal tractor oil – a combined engine/transmission oil – was realized, the first Tractor Oil Universal (TOU). This European development did not take into account the severe frictional requirements of wet brakes, which were still to appear in Europe. After 1971, the number of immersed brake systems increased rapidly, together with a corresponding volume of complaints concerning wet-brake squawk. In all haste, intense development programmes were instigated to identify friction-modifying additives with good thermal and chemical stability, suitable for use in turbocharged diesel engine oils. The resulting products, satisfying the severe technical requirements of highly turbocharged diesel engines, torque convertors, transmission gears and immersed brakes, are termed 'Super Tractor Oil Universal' (STOU). These sophisticated products meet the technical requirements of many major manufacturers, although certain suppliers of agricultural machinery still demand use of specific tractor transmission oils during the guarantee period.

22

FUTURE PERSPECTIVES

Further developments are anticipated in the direction of high-quality multi-functional fluids, permitting increased rationalization and cost savings. At the same time a requirement is envisaged for specialized products possessing high thermal and oxidation stability, capable of sustaining long service lives over wide temperature ranges. New, unconventional components with unique properties are expected to be introduced as industrial technological innovation progresses. Examples of this are already to be found in the offshore sector where, for example, hydraulic pumps and motors constructed to utilize *sea water* as the hydraulic medium are already in use (Fig. 22.1).

These components are the ideal choice for maintenance tools required by divers for sub-sea operation at considerable depths. Conventional tools are oil- or air-driven, neither being suitable for deep sea operation. Incorporation of selected polymers and corrosion-resistant metals has created interesting new components, successfully operating on this unconventional yet totally environmentally acceptable hydraulic medium. The British company pioneering these developments now markets a range of axial piston pumps and hydraulic motors suitable for water operation at pressures of 14 MPa and power ratings up to 42 kW.

The increasing use of sub-sea modules in offshore oil production, often involving satellite modules positioned at considerable distances from the production platform, has intensified the search for environmentally safe control systems. This has given rise to renewed attention towards the use of water-based fluids (ISO-type HFAS) with water contents $\leq 90\%$, capable of transmitting the required operating pres-

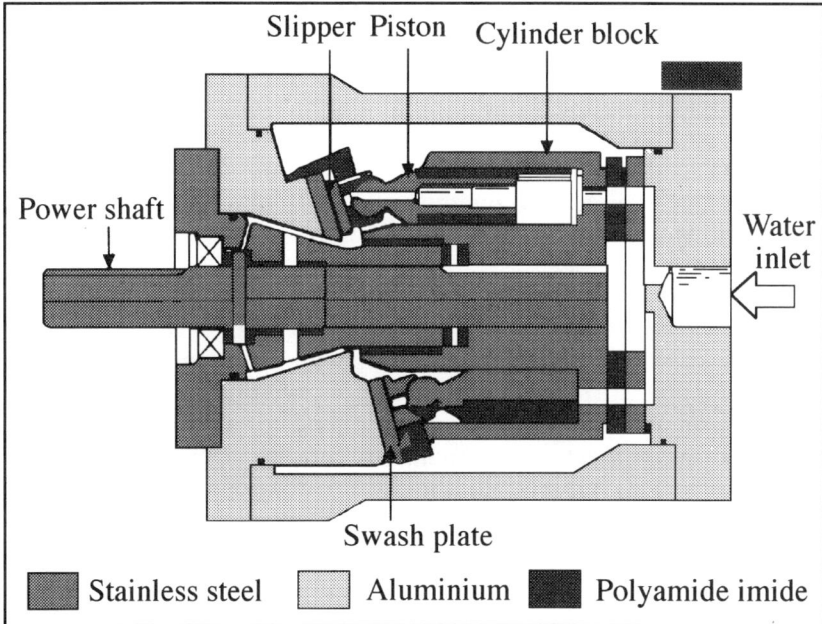

Fig. 22.1 Alternative materials for sea water operation.

sure over several kilometres at temperatures of 5–10°C. Suitable fluids must necessarily be of low viscosity, possess satisfactory lubrication properties, adequate corrosion protection, compatibility with elastomers and offer long-term protection against microbiological growth.

Offshore oil production in the North Sea has initiated developments and innovations in many directions, not least in view of the tremendous economic benefits to be gained by improved production techniques.

An interesting approach is the hydraulically driven downhole well pump developed by a Norwegian engineering company. This comprises an integrated unit of a hydraulic motor driving a peripheral centrifugal pump, the hydraulic fluid and pumped crude being transported within separate channels of concentric piping (the 'umbilical') from the production platform.

Sub-sea satellite production wells are projected with integrated units of multi-phase pumps, driven by hydraulic motors operating on high pressure water supplied via an umbilical line from a hydraulic pump unit on the operating platform. Recently developed multi-phase pumps to transport unstabilized crude oil directly from well head to land or centralized platforms represent an economically attractive alternative to established procedures when developing oil and gas fields. Crude oil is a complex mixture of oil, gas, water and also, usually, sand particles. Conventional technology requires separation of the phases prior to individual pressurization of the gas and liquid phases. Considerable cost savings are now possible when suitable equipment is available to compress the multi-phase mixture without any previous separation stage, an additional merit being the significantly simplified system.

Research directed towards improved fire-resistant hydraulic fluids for aviation, aerospace and military applications continues. In order to satisfy the most extreme performance requirements, only fluorinated compounds appear suitable at present; this seems in conflict with environmental considerations, e.g. biodegradability.

In all other respects, environmental considerations will undoubtedly receive high priority in the future development of hydraulic systems and associated functional fluids. A whole new generation of 'environmentally friendly' hydraulic fluids is expected to appear on the market during the next few years, together with a corresponding range of specialized equipment from the component manufacturers, tailored to the requirements of the new media. A competitive inducement to produce environmentally acceptable fluids has materialized in the form of various national approval schemes, i.a. *'Blau ängel'* (Germany) and 'White swan' (Scandinavia). A future harmonization of these is anticipated, e.g. under the auspices of the EU, promoting new and safer products.

23
HEALTH AND SAFETY

During recent years we have developed a better understanding of the potential hazards associated with the enormous variety of chemical substances in commercial use. Today, we are particularly aware of the need to recognize eventual health or safety hazards related to industrial and household products, and to know what precautions should be observed. Much effort is expended to ensure safe working conditions and establish a favourable environment; not only is this a question of sound common sense and ethics, but it is also strictly regulated by legislation in most countries. In the UK, the Health and Safety at Work Act places responsibility upon suppliers 'to ensure so far as is reasonably practical, that the substance is safe and without risks to health when properly used' and to make available information 'about any conditions necessary to ensure that it will be safe and without risks to health when properly used'. It is unfortunately a sober fact that no product is *completely* safe in all respects, and strict adherence to adequate safety precautions should be emphasized.

Hydraulic fluids have the advantage of being used in enclosed systems. It is therefore mainly during filling, due to leakages, or during maintenance work, that the possibility of physical contact with the hydraulic medium arises.

In general, hydraulic fluids represent a relatively insignificant health hazard in normal applications, provided relevant safety precautions in accordance with the relevant Materials Safety Data Sheet are followed during maintenance work.

During development of new formulations, the hydraulic fluid manufacturers also devote considerable effort to ensure that the occupational health aspects of the product are fully acceptable. Most countries have formal legislative requirements defining necessary hazard warning texts and symbols on the packaging of industrial products, including hydraulic fluids. In accordance with EU directives, only a few hydraulic fluids are considered to require *symbol* labelling within hazard groups Xn ('hazardous'), Xi ('irritant') etc.

Possible health hazards are summarized below.

23.1 Ingestion

The risk of swallowing hydraulic fluids would appear fairly remote, though accidents have occurred when small volumes have been refilled into mineral water bottles and stored carelessly. Any mistake will normally be registered at the first mouthful! With respect to conventional mineral hydraulic oils, ingestion of a few millilitres would not normally give rise to more than temporary discomfort, providing the fluid is *swallowed*.

The acute oral toxicity of a typical mineral oil hydraulic fluid is normally low, e.g. $LD_{50} > 5$ ml/kg.

Under no circumstances should one attempt to induce vomiting after swallowing a hydraulic fluid. Vomiting involves a risk of drawing the fluid down into the lungs ('*aspiration*'), an extremely dangerous condition which requires immediate medical attention.

Certain water-based fluids, particularly products in concentrated form, can be highly alkaline (høy pH-verdi) and highly irritant to the mouth, throat and oesophagus. When such fluids or other hazard-labelled products are swallowed, immediate medical attention should be sought.

23.2 Skin contact

The most common form of exposure to hydraulic fluids is skin contact. Mineral oils and the majority of synthetic fluids are characterized by a tendency to leach natural fats from the skin, causing dryness, irritation and eczema. Regular and prolonged skin contact may give rise to allergic reactions, and may in isolated cases lead to even more serious diseases, e.g. skin cancer.

It is consequently highly desirable to avoid unnecessary skin contact by use of suitable protective clothing and gloves of chemically resistant material, e.g. PVC. Accidental contamination of the skin by hydraulic fluid should be removed as soon as possible using soap and warm water. Special barrier creams are also available to assist in protecting the skin against contact with various types of fluids.

23.3 Eye contact

This may often feel extremely painful, even though hydraulic fluids seldom result in permanent injury to the eyes if promptly attended to. The recommended immediate treatment is to flush the eyes with copious amounts of cold water for **at least 10 minutes**. Medical attention should then be sought if any sign of irritation persists.

23.4 Inhalation

This seldom occurs to any measurable degree during normal use of hydraulic fluids, but possible vapours can irritate mucous membranes in the throat and respiratory organs. They may also cause dizziness, headaches and nausea, possibly even loss of consciousness at high concentrations. Excessively high temperatures may result in liberation of toxic decomposition products from certain products, e.g. *hydrogen sulphide* (H_2S) by thermal decomposition of dithiophosphate additives.

23.5 Materials safety data sheet

All companies should assemble a product register, including a Materials Safety Data Sheet (MSDS) for every product utilized, stored or transported by the company. This is now a formal requirement by government legislation in most industrialized countries. These data sheets should provide all necessary information concerning health hazards, handling precautions, labelling, first-aid, waste disposal, etc.

24

HYDRAULIC FLUIDS AND THE ENVIRONMENT

Latter years have been marked by increased vigilance towards the serious threats to humanity posed by industrial activities and associated pollution of the environment. Growing public interest in protecting nature – and ourselves – has materialized, and hydraulic fluids are naturally included amongst the many products against which interest is directed. Hydraulic fluids are indeed of especial interest in this respect because they are utilized in relatively large volumes in excavating machines, bulldozers, mobile cranes and other equipment for outdoor operations. The danger of leakages is always present, and the sudden rupture of a flexible hydraulic hose under pressure may in the space of a few seconds result in considerable pollution of the surroundings and ground water.

Mineral oils are composed of relatively stable hydrocarbon compounds, and are only very slowly broken down by microorganisms in the environment. Eventual pollution by conventional mineral hydraulic oils can therefore disturb the ecological balance, both in waterways and on land, for long periods. This fact has resulted in a growing interest in biodegradable products, and it is generally assumed that 'biodegradable' is synonymous with 'environmentally friendly'. However, this is not necessarily so, as many chemical compounds also form toxic biodegradation products. It is not therefore valid to equate biodegradability and environmental acceptability. Pollution of woods and hedgerows with one of today's most promising (based on current knowledge) biodegradable fluids could also result in short-term ('acute') injurious effects to trees and plants, but less long-term environmental damage than with mineral oil products.

24.1 What is biodegradability?

Microbiological degradation is the processes whereby microorganisms, with ('aerobic') or without ('anaerobic') the help of oxygen, break down organic material and extract nourishment from the decomposition products. In every gram of fertile soil there exist around 10^8 living bacteria, of an average size of $1–2$ μm^3 or, expressed in more familiar terms, 100–200 kg of microbes per acre of good agricultural land. These microorganisms multiply very rapidly when food and warmth are available.

In the *aerobic* process, microbial attack is a form of oxidation, resulting in smaller molecules of water-soluble substances capable of being utilized in the metabolism of the microorganisms. When aerobic microbiological degradation of hydrocarbons is complete, a process requiring liberal supplies of oxygen, the final products are carbon dioxide and water:

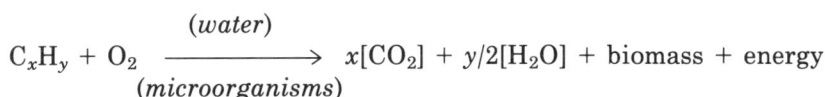

$$C_xH_y + O_2 \xrightarrow[(microorganisms)]{(water)} x[CO_2] + y/2[H_2O] + biomass + energy$$

In the alternative *anaerobic* process, which predominates deep down in the soil where oxygen availability is limited, other organic decomposition products are formed in place of carbon dioxide, e.g. methane. Both processes form part of the so-called 'carbon cycle' returning carbon compounds to the atmosphere.

The actual kinetics of the microbial processes depends upon a number of parameters, including light intensity, temperature, presence of nutrient salts, oxygen availability and type of microorganism.

However, the environmental aspect is complicated by the fact that hydraulic fluids usually contain other components, namely additives, which are often only slowly biodegraded. There would therefore appear to be grounds to query the advantages of an easily biodegradable base oil, if the additives remain as a concentrated residue in the environment. An additive concentrate would seem to represent quite a different order of environmental hazard to its original diluted form.

24.2 Determination of biodegradability

A variety of test methods have been developed, measuring for example:

- residual quantity of the original substance,
- increase in biomass (number of organisms),
- consumption of oxygen,
- formation of CO_2.

The most familiar method is CEC-L-33T-82 (issued by the Coordinating European Council), which was originally developed to investigate the biodegradability of lubricating oil for two-stroke outboard engines in water. The oil sample and microorganisms are mixed in a test flask, then stored in darkness at 25°C. The amounts of test oil remaining after 7 and 21 days are measured. The oil is considered sufficiently biodegradable if ≥80% of the organic test material is broken down within 21 days.

Another favoured method, registering the quantity of carbon dioxide formed during the test period, is the 'Modified Stürm Test'. Here the CO_2 formation is recorded as a percentage of the theoretical value for complete degradation, thus expressing the fractional decomposition, assuming microbial attack goes to completion.

A completely reliable assessment of microbial degradation and its effect on the environment can only be achieved by supplementing tests of the above nature with a detailed examination of all intermediate decomposition products, a daunting (and expensive!) task.

24.3 Biodegradable hydraulic media

Three main types of biodegradable fluids are available at present, based on:

- vegetable oils,
- polyalkylene glycols,
- organic esters.

Products formulated with *rapeseed oil* base fluid are currently most widespread, largely on account of their relatively modest cost compared to the synthetic alternatives.

Rapeseeds contain 30–40% vegetable oil. The rapeseed oil is produced by pressing followed by solvent extraction with naphtha. Rapeseed oil is principally composed of triglycerides of C_{18} to C_{22} mono-unsaturated fatty acids. The composition varies somewhat in accordance with the area of origin, and the physical and chemical properties of the products may be modified by various refining processes, e.g. hydrogenization.

The refined rapeseed oil offers a number of inherent advantages, in particular an extremely high viscosity index (>200), excellent lubrication properties, and good biodegradability. Relatively poor oxidation stability and a tendency to hydrolyse when contaminated by water are the principal factors limiting more widespread application of rapeseed oil-based fluids.

The maximum recommended temperature of operation for rapeseed

Table 24.1 Comparison of various biodegradable hydraulic fluids and ISO HV mineral oil

	Mineral oil	Rapeseed oil	Polyglycol	Ester
ISO VG	68	32	32	46
Viscosity index	145	210	200	180
Density at 15°C	0.88	0.92	1.03	0.92
Pour point (°C)	−30	−35	−39	−48
Lubrication properties	+	++	+	+
FZG value	11	12	12	10
Oxidation stability	++	+	++	++
Hydrolytic stability	++	−	+	+
Shear stability	+	++	++	++
Corrosion protection	++	+	+	+
Water solubility	No	No	Yes	No
Miscibility with mineral oil	Yes	Yes	No	Yes
Elastomer compatibility	++	+	+/−	+/−
Temperature range (°C)	−10/+90	−25/+70	−25/+100	−30/+100
Biodegradability (CEC-L-33-T-82) (%)	≈20	≈98	≈90	≈85
Wassergefährungsklasse	2	0	0	0
Cost, relative to mineral oil	1	2–3	3–4	6

oil products is 70°C, and this is too low for many of the more critical forestry machines today. At moderate temperatures, however, this type of hydraulic medium has displayed extremely satisfactory properties, and its unlimited miscibility with mineral oils ensures problem-free application in systems previously operated on conventional mineral hydraulic oils.

Polyalkylene glycols also possess good lubrication properties, a high viscosity index and – in contrast to vegetable oil products – excellent oxidation and hydrolytic stability. Water-soluble polyglycols (e.g. poly-ethylene glycols) are also rapidly biodegraded by microorganisms, and consequently this type of polyglycol is utilized in biodegradable fluids.

A significant disadvantage of polyalkylene glycols is their very limited solubility in mineral oils or rapeseed oil, this necessitating a more complicated conversion procedure for systems previously oper-ating on these fluids. Furthermore, the inherent hygroscopy of poly-alkylene glycols somewhat inhibits their ability to yield a high degree of protection against corrosion.

Synthetic esters of various types are eminently satisfactory biode-gradable base fluids for hydraulic media. Some esters possess high viscosity indices, excellent thermal and oxidation stability, attractive low temperature fluidity together with a number of other desirable features. By suitable choice of a specific ester, an optimal combination of physical and chemical properties may be obtained, e.g. satisfactory hydrolytic stability and compatibility with elastomers (a weakness with certain esters). The relatively high cost of esters tend, however, to limit wider application of these fluids.

Table 24.1 compares the properties of various typical biodegradable fluids with those of a conventional mineral hydraulic oil, ISO type HV.

Bibliography

Performance Testing of Hydraulic Fluids, Institute of Petroleum seminar, London, 1978.

van der Wulp, J.J. *Hydrauliek*, Spruyt, Van Mantgem & De Does bv, Leiden, Netherlands, 1992.

Asle, N.Y., *Handbook of Lubrication–Theory and Practice of Tribology*, CRC Press Inc., Florida, 1989.

Hydraulic-Pneumatic Symposium proceedings, Oslo. HP–Foreningen, Oslo, 1991.

Proceedings, Conference on Synthetic Lubricants, Sopron. Hungarian Hydrocarbon Institute, Százhalombatta, 1989.

Onsøyen, E., *Reliability–Maintenance* (in Norwegian). Lecture notes, Norway's Technical University, Trondheim, 1991.

Hatton, D.R, *Some Practical Aspects of Hydraulic Fluids*, Technical paper, Shell International, London, 1989.

The Hydraulic Trainer, Rexroth GMBH, Germany, 1978.

Jackson, T.L., *Selection of Materials and Fluids for Use in Hydraulic Systems*, M.O.D./I.Mar.E. joint symposium, London, 1973.

Institute of Petroleum, London, *Modern Petroleum Technology*, John Wiley & Sons, 1984.

APPENDIX 1

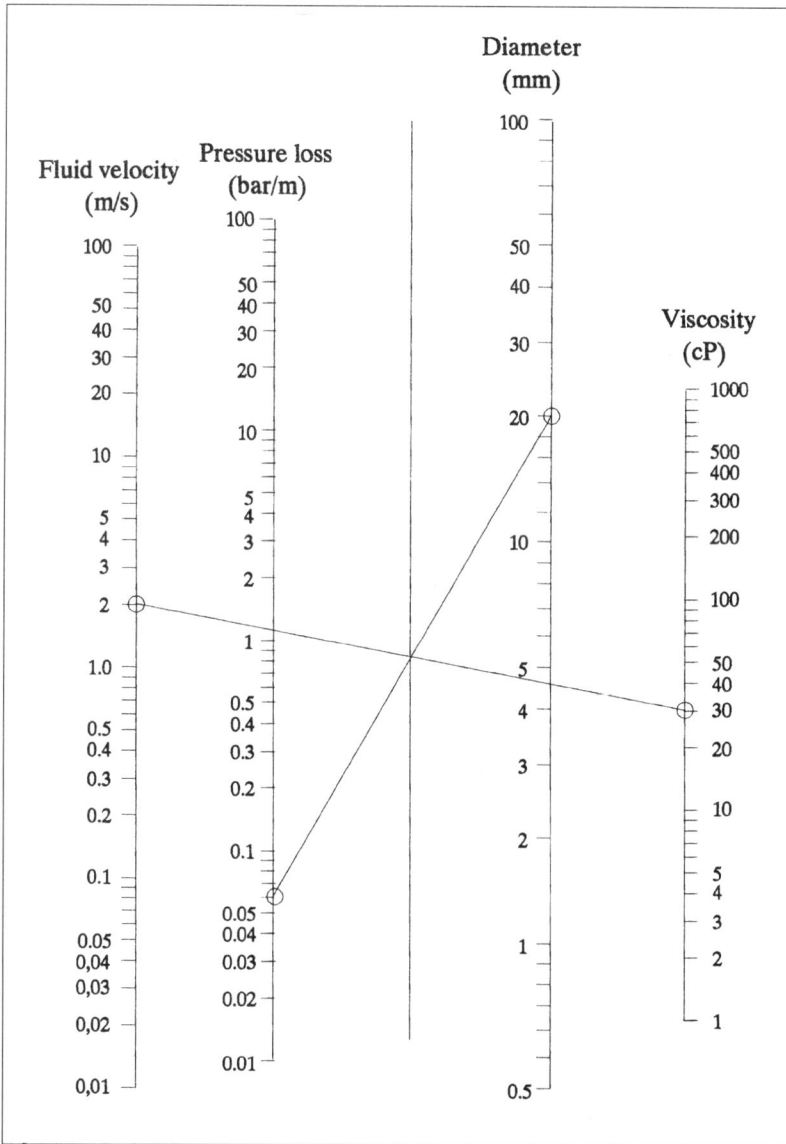

Fig. A1 Pressure losses per unit length in pipes (laminar flow, Re < 2000).

APPENDIX 2

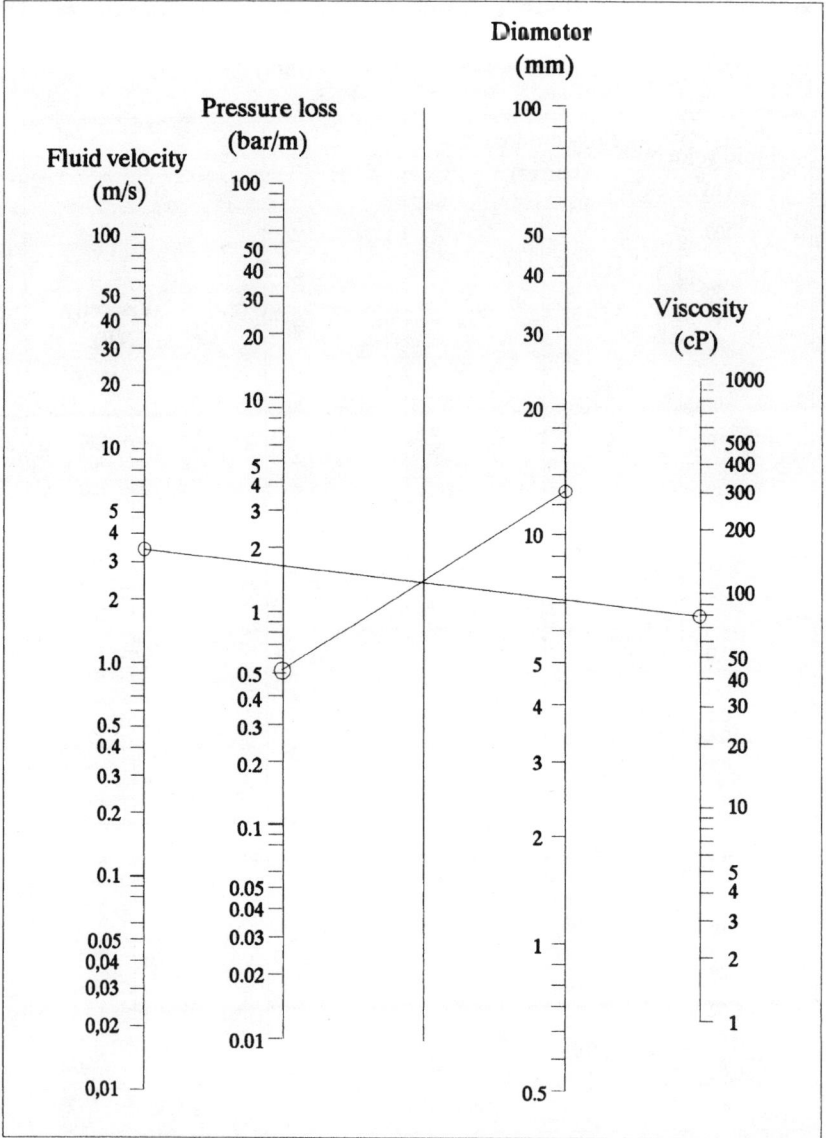

Fig. A2 Pressure losses per unit length in pipes (turbulent flow Re > 2000).

INDEX